Intelligent Systems Reference Library

Volume 69

For further volumes:
http://www.springer.com/series/8578

About this Series

The aim of this series is to publish a Reference Library, including novel advances and developments in all aspects of Intelligent Systems in an easily accessible and well structured form. The series includes reference works, handbooks, compendia, textbooks, well-structured monographs, dictionaries, and encyclopedias. It contains well integrated knowledge and current information in the field of Intelligent Systems. The series covers the theory, applications, and design methods of Intelligent Systems. Virtually all disciplines such as engineering, computer science, avionics, business, e-commerce, environment, healthcare, physics and life science are included.

Catalin Stoean · Ruxandra Stoean

Support Vector Machines and Evolutionary Algorithms for Classification

Single or Together?

Catalin Stoean
Faculty of Mathematics and Natural Sciences
Department of Computer Science
University of Craiova
Craiova
Romania

Ruxandra Stoean
Faculty of Mathematics and Natural Sciences
Department of Computer Science
University of Craiova
Craiova
Romania

ISSN 1868-4394 ISSN 1868-4408 (electronic)
ISBN 978-3-319-38243-2 ISBN 978-3-319-06941-8 (eBook)
DOI 10.1007/978-3-319-06941-8
Springer Cham Heidelberg New York Dordrecht London

Softcover reprint of the hardcover 1st edition 2014

Printed on acid-free paper

Springer is part of Springer Science+Business Media (www.springer.com)

To our sons, Calin and Radu

Foreword

Indisputably, *Support Vector Machines* (SVM) and *Evolutionary Algorithms* (EA) are both established algorithmic techniques and both have their merits and success stories. It appears natural to combine the two, especially in the context of classification. Indeed, many researchers have attempted to bring them together in or or the other way. But if I would be asked who could deliver the most complete coverage of all the important aspects of interaction between SVMs and EAs, together with a thorough introduction into the individual foundations, the authors would be my first choice, the most suitable candidates for this endeavor.

It is now more than ten years ago that I first met Ruxandra, and almost ten years since I first met Catalin, and we have shared a lot of exciting research related and more personal (but not less exciting) moments, and more is yet to come, as I hope. Together, we have experienced some cool scientific successes and also a bitter defeat when somebody had the same striking idea on one aspect of SVM and EA combination and published the paper when we had just generated the first, very encouraging experimental results. The idea was not bad, nonetheless, because the paper we did not write won a best paper award.

Catalin and Ruxandra are experts in SVMs and EAs, and they provide more than an overview over the research on the combination of both with a focus on their own contributions: they also point to interesting interactions that desire even more investigation. And, unsurprisingly, they manage to explain the matter in a way that makes the book very approachable and fascinating for researchers or even students who only know one of the fields, or are completely new to both of them.

Bochum, February 2014 Mike Preuss

Preface

When we decided to write this book, we asked ourselves whether we could try and unify everything that we have studied and developed under a same roof, where a reader could find some of the old and the new, some of the questions and several likely answers, some of the theory around support vector machines and some of the practicality of evolutionary algorithms. All working towards a common target: classification. We use it everyday, even without being aware of it: we categorize people, food, music, movies, books. But when classification is involved at a larger scale, like for the provision of living, health and security, effective computational means to address it are vital.

This work, describing some of its facets in connection to support vector machines and evolutionary algorithms, is thus an appropriate reading material for researchers in machine learning and data mining with an emphasis on evolutionary computation and support vector learning for classification. The basic concepts and the literature review are however suitable also for introducing MSc and PhD students to these two fields of computational intelligence. The book should be also interesting for the practical environment, with an accent on computer aided diagnosis in medicine. Physicians and those working in designing computational tools for medical diagnosis will find the discussed techniques helpful, as algorithms and experimental discussions are included in the presentation.

There are many people who are somehow involved in the emergence of this book. We thank Dr. Camelia Pintea for convincing and supporting us to have it published. We express our gratitude to Prof. Lakhmi Jain, who so warmly sustained this project. Acknowledgements also go to Dr. Thomas Ditzinger, who so kindly agreed to its appearance.

Many thanks to Dr. Mike Preuss, who has been our friend and co-author for so many years now; from him we have learnt how to experiment thoroughly and how to write convincingly. We are also grateful to Prof. Thomas Bartz-Beielstein, who has shown us friendship and the SPO. We also thank him, as well as Dr. Boris Naujoks and Martin Zaefferer, for taking the time to review this book before being published. Further on, without the continuous aid of Prof. Hans-Paul Schwefel and Prof. Günter Rudolph, we would not have started and continued our fruitful collaboration with our German research partners; thanks also to the nice staff at TU Dortmund and FH Cologne. In the same sense, we owe a lot to the Deutscher Akademischer Austauschdienst (DAAD) who supported our several research stays in Germany. Our thoughts go as well to Prof. D. Dumitrescu, who introduced us to evolutionary algorithms and support vector machines and who has constantly encouraged us, all throughout PhD and beyond, to push the limits in our research work and dreams.

We also acknowledge that this work was partially supported by the grant number 42C/2014, awarded in the internal grant competition of the University of Craiova. We also thank our colleagues from its Department of Computer Science for always stimulating our research.

Our families deserve a lot of appreciation for always being there for us. And last but most importantly, our love goes to our sons, Calin and Radu; without them, we would not have written this book with such optimism, although we would have finished it faster. Now, that it is complete, we will have more time to play together. Although our almost 4-year old son solemnly just announced us that we would have to defer playing until he also finished writing his own book.

Craiova, Romania
March 2014

Catalin Stoean
Ruxandra Stoean

Contents

Acronyms

SVM	Support vector machine
PP	Primal problem
SRM	Structural risk minimization
VC	Vapnik-Chervonenkis
KKTL	Karush-Kuhn-Tucker-Lagrange
DP	Dual problem
EA	Evolutionary algorithm
EC	Evolutionary computation
GA	Genetic algorithm
GC	Genetic chromodynamics
CC	Cooperative coevolution
ESVM	Evolutionary-driven support vector machine
SVM-CC	Support vector machines followed by cooperative coevolution
HC	Hill climbing
DT	Decision trees
NN	Neural network
SPO	Sequential parameter optimization
LHS	Latin hypercube sampling
UCI	University of California at Irvine

Chapter 1
Introduction

> ***The beginning is the most important part of the work.***
> *Plato, The Republic*

Suppose one is confronted with a medical classification problem. What trustworthy technique should one then use to solve it? Support vector machines (SVMs) are known to be a smart choice. But how can one make a personal, more flexible implementation of the learning engine that makes them run that well? And how does one open the black box behind their predicted diagnosis and explain the reasoning to the otherwise reluctant fellow physicians? Alternatively, one could choose to develop a more versatile evolutionary algorithm (EA) to tackle the classification task towards a potentially more understandable logic of discrimination. But will comprehensibility weigh more than accuracy?

It is therefore the goal of this book to investigate how can both efficiency as well as transparency in prediction be achieved when dealing with classification by means of SVMs and EAs. We will in turn address the following choices:

1. Proficient, black box SVMs (found in chapter 2).
2. Transparent but less efficient EAs (chapters 3, 4 and 5).
3. Efficient learning by SVMs, flexible training by EAs (chapter 6).
4. Predicting by SVMs, explaining by EAs (chapter 7).

The book starts by reviewing the classical as well as the state of the art approaches to SVMs and EAs for classification, as well as methods for their hybridization. Nevertheless, it is especially focused on the authors' personal contributions to the enunciated scope.

Each presented new methodology is accompanied by a short experimental section on several benchmark data sets to get a grasp of its results. For more in-depth experimentally-related information, evaluation and test cases the reader should consult the corresponding referenced articles.

Throughout this book, we will assume that a classification problem is defined by the subsequent components:

- a set of m training pairs, where each holds the information related to a data sample (a sequence of values for given attributes or indicators) and its confirmed target (outcome, decision attribute).

C. Stoean and R. Stoean, *Support Vector Machines and Evolutionary Algorithms for Classification*, Intelligent Systems Reference Library 69,
DOI: 10.1007/978-3-319-06941-8_1, © Springer International Publishing Switzerland 2014

- every sample (or example, record, point, instance) is described by n attributes: $x_i \in [a_1,b_1] \times [a_2,b_2] \times ... \times [a_n,b_n]$, where a_i,b_i denote the bounds of definition for every attribute.
- each corresponding outcome $y_i \in \{0,1,...,k-1\}$, where there are k possible classes.
- a set of l validation couples (x_i^v, y_i^v), in order to assess the prediction error of the model. Please note that this set can be constituted only in the situation when the amount of data is sufficiently large [Hastie et al, 2001].
- a set of p test pairs of the type $(x_i^{'}, y_i^{'})$, to measure the generalization error of the approach [Hastie et al, 2001].
- for both the validation and test sets, the target is unknown to the learning machine and must be predicted.

As illustrated in Fig. 1.1, learning pursues the following steps:

- A chosen classifier learns the associations between each training sample and the acknowledged output (training phase).
- Either in a black box manner or explicitly, the obtained inference engine takes each test sample and makes a forecast on its probable class, according to what has been learnt (testing phase).
- The percent of correctly labeled new cases out of the total number of test samples is next computed (accuracy of prediction).
- Cross-validation (as in statistics) must be employed in order to estimate the prediction accuracy that the model will exhibit in practice. This is done by selecting training/test sets for a number of times according to several possible schemes.
- The generalization ability of the technique is eventually assessed by computing the test prediction accuracy as averaged over the several rounds of cross-validation.
- Once more, if we dispose of a substantial data collection, it is advisable to additionally make a prediction on the targets of validation examples, prior to the testing phase. This allows for an estimation of the prediction error of the constructed model, computed also after several rounds of cross-validation that now additionally include the validation set [Hastie et al, 2001].

Note that, in all conducted experiments throughout this book, we were not able to use the supplementary validation set, since the data samples in the chosen sets were insufficient. This was so because, for the benchmark data sets, we selected those that were both easier to understand for the reader and cleaner to make reproducing of results undemanding. For the real-world available tasks, the data was not too numerous as it comes from hospitals in Romania, where such sets have been only recently collected and prepared for computer-aided diagnosis purposes.

What is more, we employ the repeated random sub-sampling method for cross-validation, where the multiple training/test sets are chosen by randomly splitting the data in two for the given number of times.

As the task for classification is to achieve an optimal separation of given data into classes, SVMs regard learning from a geometrical point of view. They assume the existence of a separating surface between every two classes labeled as -1 and

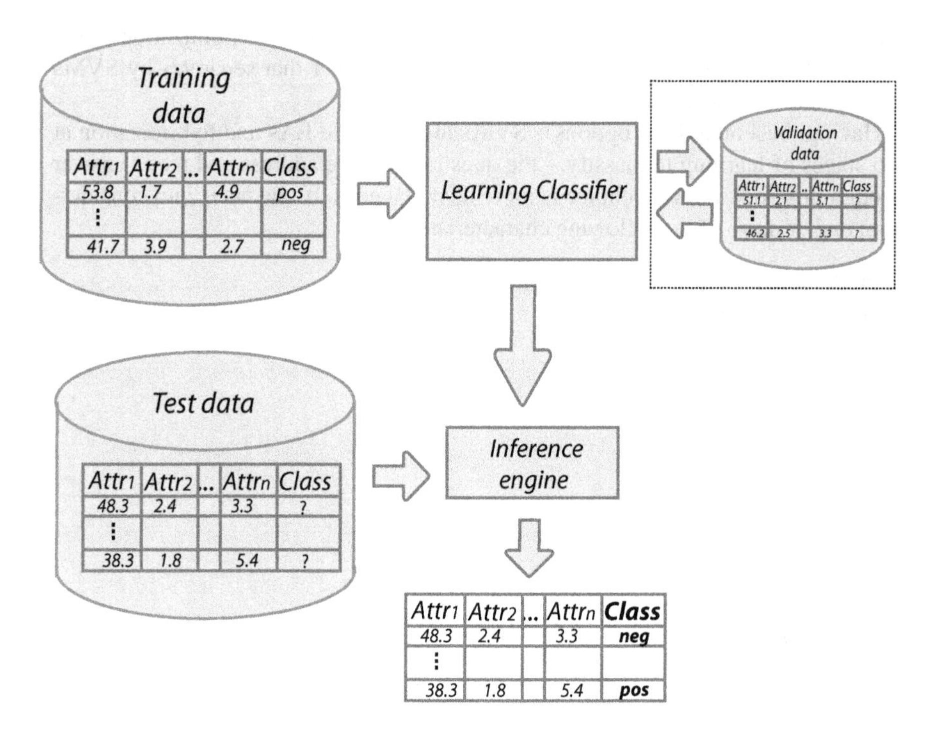

Fig. 1.1 The classifier learns the associations between the training samples and their corresponding classes and is then calibrated on the validation samples. The resulting inference engine is subsequently used to classify new test data. The validation process can be omitted, especially for relatively small data sets. The process is subject to cross-validation, in order to estimate the practical prediction accuracy.

1. The aim then becomes the discovery of the appropriate decision hyperplane. The book will outline all the aspects related to classification by SVMs, including the theoretical background and detailed demonstrations of their behavior (chapter 2).

EAs, on the other hand, are able to evolve rules that place each sample into a corresponding class, while training on the available data. The rules can take different forms, from the IF-THEN conjunctive layout from computational logic to complex structures like trees. In this book, we will evolve thresholds for the attributes of the given data examples. These IF-THEN constructions can also be called rules, but we will more rigorously refer to them as class prototypes, since the former are generally supposed to have a more elaborate formulation. Two techniques that evolve class prototypes while maintaining diversity during evolution are proposed: a multimodal EA that separates potential rules of different classes through a common radius means (chapter 4) and another that creates separate collaborative populations connected to each outcome (chapter 5).

Combinations between SVMs and EAs have been widely explored by the machine learning community and on different levels. Within this framework, we outline approaches tackling two degrees of hybridization: EA optimization at the core of SVM learning (chapter 6) and a stepwise learner that separates by SVMs and explains by EAs (chapter 7).

Having presented these options – SVMs alone, single EAs and hybridization at two stages of learning to classify – the question that we address and try to answer through this book is: what choice is more advantageous, if one takes into consideration one or more of the following characteristics:

- prediction accuracy
- comprehensibility
- simplicity
- flexibility
- runtime

Part I
Support Vector Machines

The first part of this book describes support vector machines from (a) their geometrical view upon learning to (b) the standard solving of their inner resulting optimization problem. All the important concepts and deductions are thoroughly outlined, all because SVMs are very popular but most of the time not understood.

Chapter 2
Support Vector Learning and Optimization

East is east and west is west and never the twain shall meet.
The Ballad of East and West by Rudyard Kipling

2.1 Goals of This Chapter

The kernel-based methodology of SVMs [Vapnik and Chervonenkis, 1974], [Vapnik, 1995a] has been established as a top ranking approach for supervised learning within both the theoretical and red practical research environments. This very performing technique suffers nevertheless from the curse of an opaque engine [Huysmans et al, 2006], which is undesirable for both theoreticians, who are keen to control the modeling, and the practitioners, who are more than often suspicious of using the prediction results as a reliable assistant in decision making.

A concise view on a SVM is given in [Cristianini and Shawe-Taylor, 2000]:

> A system for efficiently training linear learning machines in kernel-induced feature spaces, while respecting the insights of generalization theory and exploiting optimization theory.

The right placement of data samples to be classified triggers corresponding separating surfaces within SVM training. The technique basically considers only the general case of binary classification and treats reductions of multi-class tasks to the former. We will also start from the general case of two-class problems and end with the solution to several classes.

If the first aim of this chapter is to outline the essence of SVMs, the second one targets the presentation of what is often presumed to be evident and treated very rapidly in other works. We therefore additionally detail the theoretical aspects and mechanism of the classical approach to solving the constrained optimization problem within SVMs.

Starting from the central principle underlying the paradigm (Sect. 2.2), the discussion of this chapter pursues SVMs from the existence of a linear decision function (Sect. 2.3) to the creation of a nonlinear surface (Sect. 2.4) and ends with the treatment for multi-class problems (Sect. 2.5).

C. Stoean and R. Stoean, *Support Vector Machines and Evolutionary Algorithms for Classification*, Intelligent Systems Reference Library 69,
DOI: 10.1007/978-3-319-06941-8_2, © Springer International Publishing Switzerland 2014

2.2 Structural Risk Minimization

SVMs act upon a fundamental theoretical assumption, called the principle of structural risk minimization (SRM) [Vapnik and Chervonenkis, 1968].

Intuitively speaking, the SRM principle asserts that, for a given classification task, with a certain amount of training data, generalization performance is solely achieved if the accuracy on the particular training set and the capacity of the machine to pursue learning on any other training set without error have a good balance. This request can be illustrated by the example found in [Burges, 1998]:

> A machine with too much capacity is like a botanist with photographic memory who, when presented with a new tree, concludes that it is not a tree because it has a different number of leaves from anything she has seen before; a machine with too little capacity is like the botanist's lazy brother, who declares that if it's green, then it's a tree. Neither can generalize well.

We have given a definition of classification in the introductory chapter and we first consider the case of a binary task. For convenience of mathematical interpretation, the two classes are labeled as -1 and 1; henceforth, $y_i \in \{-1,1\}$.

Let us suppose the set of functions $\{f_t\}$, of generic parameters t:

$$f_t : \mathbb{R}^n \rightarrow \{-1,1\}. \tag{2.1}$$

The given set of m training samples can be labeled in 2^m possible ways. If for each labeling, a member of the set $\{f_t\}$ can be found to correctly assign those labels, then it is said that the collection of samples is shattered by that set of functions [Cherkassky and Mulier, 2007].

Definition 2.1. [Burges, 1998] The Vapnik-Chervonenkis (VC) - dimension h for a set of functions $\{f_t\}$ is defined as the maximum number of training samples that can be shattered by it.

Proposition 2.1. *(Structural Risk Minimization principle) [Vapnik, 1982]*

For the considered classification problem, for any generic parameters t and for $m > h$, with a probability of at least $1-\eta$, the following inequality holds:

$$R(t) \leq R_{emp}(t) + \phi\left(\frac{h}{m}, \frac{\log(\eta)}{m}\right),$$

where R(t) is the test error, $R_{emp}(t)$ is the training error and ϕ is called the confidence term and is defined as:

$$\phi\left(\frac{h}{m}, \frac{\log(\eta)}{m}\right) = \sqrt{\frac{h\left(\log\frac{2m}{h}+1\right) - \log\frac{\eta}{4}}{m}}.$$

The SRM principle affirms that, for a high generalization ability, both the training error and the confidence term must be kept minimal; the latter is minimized by reducing the VC-dimension.

2.3 Support Vector Machines with Linear Learning

When confronted with a new classification task, the first reasonable choice is to try and separate the data in a linear fashion.

2.3.1 Linearly Separable Data

If training data are presumed to be linearly separable, then there exists a linear hyperplane H:

$$H : w \cdot x - b = 0, \tag{2.2}$$

which separates the samples according to their classes [Haykin, 1999]. w is called the weight vector and b is referred to as the bias.

Recall that the two classes are labeled as -1 and 1. The data samples of class 1 thus lie on the positive side of the hyperplane and their negative counterparts on the opposite side.

Proposition 2.2. *[Haykin, 1999]*

Two subsets of n-dimensional samples are linearly separable iff there exist $w \in \mathbb{R}^n$ and $b \in \mathbb{R}$ such that for every sample $i = 1,2,...,m$:

$$\begin{cases} w \cdot x_i - b > 0, y_i = 1 \\ w \cdot x_i - b \leq 0, y_i = -1 \end{cases} \tag{2.3}$$

An insightful picture of this geometric separation is given in Fig. 2.1.

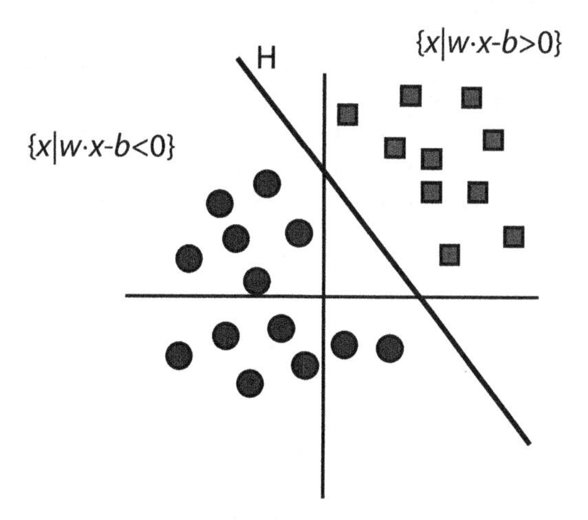

Fig. 2.1 The positive and negative samples, denoted by squares and circles, respectively. The decision hyperplane between the two corresponding separable subsets is H.

It is further resorted to a stronger statement for linear separability, where the positive and negative samples lie behind a corresponding supporting hyperplane.

Proposition 2.3. *[Bosch and Smith, 1998] Two subsets of n-dimensional samples are linearly separable iff there exist $w \in \mathbb{R}^n$ and $b \in \mathbb{R}$ such that, for every sample $i = 1, 2, ..., m$:*

$$\begin{cases} w \cdot x_i - b \geq 1, y_i = 1 \\ w \cdot x_i - b \leq -1, y_i = -1 \end{cases} \tag{2.4}$$

An example for the stronger separation concept is given in Fig. 2.2.

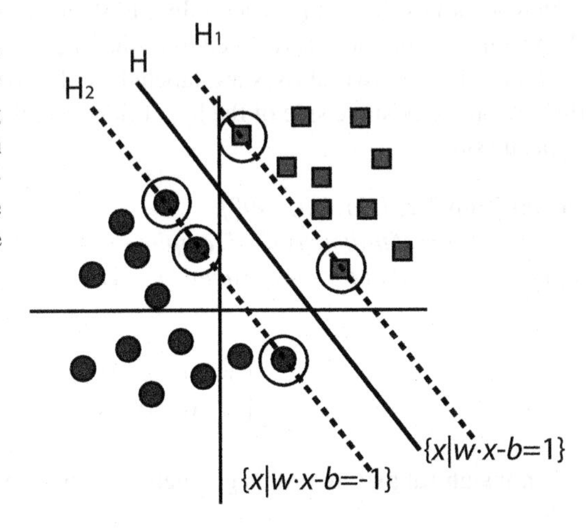

Fig. 2.2 The decision and supporting hyperplanes for the linearly separable subsets. The separating hyperplane H is the one that lies in the middle of the two parallel supporting hyperplanes H_1, H_2 for the two classes. The support vectors are circled.

Proof. (we provide a detailed version – as in [Stoean, 2008] – for a gentler flow of the connections between the different conceptual statements)

Suppose there exist w and b such that the two inequalities hold.

The subsets given by $y_i = 1$ and $y_i = -1$, respectively, are linearly separable since all positive samples lie on one side of the hyperplane given by

$$w \cdot x - b = 0,$$

from:

$$w \cdot x_i - b \geq 1 > 0 \text{ for } y_i = 1,$$

and simultaneously:

$$w \cdot x_i - b \leq -1 < 0 \text{ for } y_i = -1,$$

so all negative samples lie on the other side of this hyperplane.

Now, conversely, suppose the two subsets are linearly separable. Then, there exist $w \in \mathbb{R}^n$ and $b \in \mathbb{R}$ such that, for $i = 1, 2, ..., m$:

$$\begin{cases} w \cdot x_i - b > 0, y_i = 1 \\ w \cdot x_i - b \leq 0, y_i = -1 \end{cases}$$

Since:

$$\min\{w \cdot x_i | y_i = 1\} > \max\{w \cdot x_i | y_i = -1\},$$

let us set:

$$p = \min\{w \cdot x_i | y_i = 1\} - \max\{w \cdot x_i | y_i = -1\}$$

and make:

$$w' = \frac{2}{p} w$$

and

$$b' = \tfrac{1}{p} (\min\{w \cdot x_i | y_i = 1\} + \max\{w \cdot x_i | y_i = -1\})$$

Then:

$$\begin{aligned} \min\left\{w' \cdot x_i | y_i = 1\right\} = & \\ & = \frac{2}{p} \min\{w \cdot x_i | y_i = 1\} \\ & = \frac{1}{p} (\min\{w \cdot x_i | y_i = 1\} + \max\{w \cdot x_i | y_i = -1\} + \\ & \min\{w \cdot x_i | y_i = 1\} - \max\{w \cdot x_i | y_i = -1\}) \\ & = \frac{1}{p} (\min\{w \cdot x_i | y_i = 1\} + \max\{w \cdot x_i | y_i = -1\} + p) \\ & = b' + 1 \end{aligned}$$

and

$$\begin{aligned} \max\left\{w' \cdot x_i | y_i = -1\right\} = & \\ & = \frac{2}{p} \max\{w \cdot x_i | y_i = -1\} \\ & = \frac{1}{p} (\min\{w \cdot x_i | y_i = 1\} + \max\{w \cdot x_i | y_i = -1\} - p) \\ & = b' - 1 \end{aligned}$$

Consequently, there exist $w \in \mathbb{R}^n$ and $b \in \mathbb{R}$ such that:

$$w \cdot x_i \geq b + 1 \Rightarrow w \cdot x_i - b \geq 1 \text{ when } y_i = 1$$

$$\text{and } w \cdot x_i \leq b - 1 \Rightarrow w \cdot x_i - b \leq -1 \text{ when } y_i = -1$$

□

Definition 2.2. The support vectors are the training samples for which either the first or the second line of (2.4) holds with the equality sign.

In other words, the support vectors are the data samples that lie closest to the decision surface. Their removal would change the found solution. The supporting hyperplanes are those denoted by the two lines in (2.4), if equalities are stated instead.

Following the geometrical separation statement (2.4), SVMs hence have to determine the optimal values for the coefficients w and b of the decision hyperplane that linearly partitions the training data. In a more succinct formulation, from (2.4), the optimal w and b must then satisfy for every $i = 1, 2, ..., m$:

$$y_i(w \cdot x_i - b) - 1 \geq 0 \tag{2.5}$$

In addition, according to the SRM principle (Proposition 2.1), separation must be performed with a high generalization capacity. In order to also address this point, in the next lines, we will first calculate the margin of separation between classes.

The distance from one random sample z to the separating hyperplane is given by:

$$\frac{|w \cdot z - b|}{||w||}. \tag{2.6}$$

Let us subsequently compute the same distance from the samples z_i that lie closest to the separating hyperplane on either side of it (the support vectors, see Fig. 2.2). Since z_i are situated closest to the decision hyperplane, it results that either $z_i \in H_1$ or $z_i \in H_2$ (according to Def. 2.2) and thus $|w \cdot z_i - b| = 1$, for all i.

Hence:

$$\frac{|w \cdot z_i - b|}{||w||} = \frac{1}{||w||} \text{ for all } i = 1, 2, ..., m. \tag{2.7}$$

Then, the margin of separation becomes equal to [Vapnik, 2003]:

$$\frac{2}{||w||}. \tag{2.8}$$

Proposition 2.4. *[Vapnik, 1995b]*

Let r be the radius of the smallest ball

$$B_r(a) = \{x \in \mathbb{R}^n | \, ||x - a|| < r\}, a \in \mathbb{R}^n$$

containing the samples $x_1, ..., x_m$ and let

$$f_{w,b} = sgn(w \cdot x - b)$$

be the hyperplane decision functions.

Then the set $\{f_{w,b} | \, ||w|| \leq A\}$ has a VC-dimension h (as from Definition 2.1) satisfying

$$h < r^2 A^2 + 1$$

In other words, it is stated that, since $||w||$ is inversely proportional to the margin of separation (from (2.8)), by requiring a large margin (i.e., a small A), a small VC-dimension is obtained. Conversely, by allowing separations with small margin, a much larger class of problems can be potentially separated (i.e., there exists a larger class of possible labeling modes for the training samples, from the definition of the VC-dimension).

The SRM principle requests that, in order to achieve high generalization of the classifier, training error and VC-dimension must be both kept small. Therefore, hyperplane decision functions must be constrained to maximize the margin, i.e.,

$$\text{minimize } \frac{||w||^2}{2}, \tag{2.9}$$

and separate the training data with as few exceptions as possible.

From (2.5) and (2.9), it follows that the resulting optimization problem is (2.10) [Haykin, 1999]:

$$\begin{cases} \text{find } w \text{ and } b \text{ as to minimize } \dfrac{||w||^2}{2} \\ \text{subject to } y_i(w \cdot x_i - b) \geq 1, \text{ for all } i = 1, 2, ..., m \end{cases} \tag{2.10}$$

The reached constrained optimization problem is called the primal problem (PP).

2.3.2 *Solving the Primal Problem*

The original solving of the PP (2.10) requires the a priori knowledge of several fundamental mathematical propositions described in the subsequent lines.

Definition 2.3. A function $f: C \to \mathbb{R}$ is said to be convex if
$f(\alpha x + (1-\alpha)y) \leq \alpha f(x) + (1-\alpha)f(y)$, for all $x, y \in C$ and $\alpha \in [0,1]$.

Proposition 2.5. *For a function $f: (a,b) \to \mathbb{R}$, $(a,b) \subseteq \mathbb{R}$, that has a second derivative in (a,b), a necessary and sufficient condition for its convexity on that interval is that the second derivative $f''(x) \geq 0$, for all $x \in (a,b)$.*

Proposition 2.6. *If two functions are convex, the composition of the functions is convex.*

Proposition 2.7. *The objective function in PP* (2.10) *is convex [Haykin, 1999].*

Proof. (detailed as in [Stoean, 2008])

Let $h = f \circ g$, where $f: \mathbb{R} \to \mathbb{R}$, $f(x) = x^2$ and $g: \mathbb{R}^n \to \mathbb{R}$, $g(w) = ||w||$.

1. $f: \mathbb{R} \to \mathbb{R}$, $f(x) = x^2 \Rightarrow f'(x) = 2x \Rightarrow f''(x) = 2 \geq 0 \Rightarrow f$ is convex.

2. $g : \mathbb{R}^n \rightarrow \mathbb{R}$, $g(w) = \|w\|$

We appeal to two well-known properties of a norm:

1. $\|\alpha v\| = |\alpha| \, \|v\|$
2. $\|v + w\| \leq \|v\| + \|w\|$

Let $v, w \in \mathbb{R}^n$ and $\alpha \in [0,1]$.

$$g(\alpha v + (1-\alpha)w) = \|\alpha v + (1-\alpha)w\| \leq |\alpha| \, \|v\| + |1-\alpha| \, \|w\| = \alpha \|v\| + (1-\alpha) \|w\| = \alpha g(v) + (1-\alpha) g(w)$$

$\Rightarrow g$ is convex.

Following Proposition 2.6 $\Rightarrow h$ is convex. □

Since constraints in PP (2.10) are linear in w, the following proposition arises.

Proposition 2.8. *The feasible region for a constrained optimization problem is convex if the constraints are linear.*

At this point, we have all the necessary information to outline the classical solving of the PP inside SVMs (2.10). The standard method of finding the optimal solution with respect to the defined constraints resorts to an extension of the Lagrange multipliers method. This is described in detail in what follows.

Since the objective function is convex and constraints are linear, the Karush-Kuhn-Tucker-Lagrange (KKTL) conditions can be stated for PP [Haykin, 1999] .

This is based on the argument that, since constraints are linear, the KKTL conditions are guaranteed to be necessary. Also, since PP is convex (convex objective function + convex feasible region), the KKTL conditions are at the same time sufficient for global optimality [Fletcher, 1987].

First, the Lagrangian function is constructed:

$$L(w,b,\alpha) = \frac{1}{2}\|w\|^2 - \sum_{i=1}^{m} \alpha_i \left[y_i(w \cdot x_i - b) - 1\right], \tag{2.11}$$

where variables $\alpha_i \geq 0$ are the Lagrange multipliers.

The solution to the problem is determined by the KKTL conditions for every sample $i = 1, 2, ..., m$ [Burges, 1998]:

$$\begin{cases} \dfrac{\partial L(w,b,\alpha)}{\partial w} = 0 \\ \dfrac{\partial L(w,b,\alpha)}{\partial b} = 0 \\ \alpha_i \left[y_i(w \cdot x_i - b) - 1\right] = 0 \end{cases}$$

Application of the KKTL conditions yields [Haykin, 1999]:

$$\frac{\partial L(w,b,\alpha)}{\partial w} = w - \sum_{i=1}^{m} \alpha_i y_i x_i = 0 \Rightarrow w = \sum_{i=1}^{m} \alpha_i y_i x_i \tag{2.12}$$

$$\frac{\partial L(w,b,\alpha)}{\partial b} = \sum_{i=1}^{m} \alpha_i y_i = 0 \tag{2.13}$$

$$\alpha_i \left[y_i(w \cdot x_i - b) - 1\right] = 0, i = 1,2,...,m \tag{2.14}$$

We additionally refer to the separability statement and the conditions for positive Lagrange multipliers for every $i = 1,2,...,m$:

$$y_i(w \cdot x_i - b) - 1 \geq 0$$

$$\alpha_i \geq 0$$

We have to solve the particular PP in (2.10). Generally speaking, given the PP:

$$\begin{cases} \text{minimize } f(x) \\ \text{subject to } \begin{cases} g_1(x) \geq 0 \\ \ldots \\ g_m(x) \geq 0 \end{cases}, \end{cases} \tag{2.15}$$

the Lagrange multipliers are $\alpha = (\alpha_1^*, ..., \alpha_m^*)$, $\alpha_i^* \geq 0$, such that:

$$\inf_{g_1(x)\geq 0,...,g_m(x)\geq 0} f(x) = \inf_{x\in\mathbb{R}^n} L(x,\alpha^*),$$

where L is the Lagrangian function:

$$L(x,\alpha) = f(x) + \sum_{j=1}^{m} \alpha_j g_j(x), x \in \mathbb{R}^n, \alpha \in \mathbb{R}^m$$

Then, one can resort to the dual function [Haykin, 1999]:

$$q(\alpha) = \inf_{x\in\mathbb{R}^n} L(x,\alpha)$$

This naturally leads to the dual problem (DP) :

$$\begin{cases} \text{maximize } q(\alpha) \\ \text{subject to } \alpha \geq 0 \end{cases} \tag{2.16}$$

The optimal primal value is $f^* = \inf_{g_1(x)\geq 0,...,g_r(x)\geq 0} f(x) = \inf_{x\in\mathbb{R}^n} \sup_{\alpha\geq 0} L(x,\alpha)$.

The optimal dual value is $g^* = \sup_{\alpha\geq 0} q(\alpha) = \sup_{\alpha\geq 0} \inf_{x\in\mathbb{R}^n} L(x,\alpha)$.

There is always that $q^* \leq f^*$.

But, if there is convexity in the PP, then:

1. $q^* = f^*$
2. Optimal solutions of the DP are multipliers for the PP.

Further on, (2.11) is expanded and one obtains [Haykin, 1999]:

$$L(w,b,\alpha) = \frac{1}{2}\|w\|^2 - \sum_{i=1}^{m} \alpha_i y_i w \cdot x_i + b \sum_{i=1}^{m} \alpha_i y_i + \sum_{i=1}^{m} \alpha_i \tag{2.17}$$

The third term on the right-hand side of the expansion is zero from (2.13). Moreover, from (2.12), one obtains:

$$\frac{1}{2}\|w\|^2 = w \cdot w = \sum_{i=1}^{m} \alpha_i y_i w \cdot x_i = \sum_{i=1}^{m}\sum_{j=1}^{m} \alpha_i \alpha_j y_i y_j x_i \cdot x_j$$

Therefore, (2.17) changes to:

$$L(w,b,\alpha) = \sum_{i=1}^{m} \alpha_i - \frac{1}{2}\sum_{i=1}^{m}\sum_{j=1}^{m} \alpha_i \alpha_j y_i y_j x_i \cdot x_j$$

According to the duality concepts, by setting $Q(\alpha) = L(w,b,\alpha)$, one obtains the DP:

$$\begin{cases} \text{find } \{\alpha_i\}_{i=1,2,\ldots,m} \text{ as to maximize } Q(\alpha) = \sum_{i=1}^{m} \alpha_i - \frac{1}{2}\sum_{i=1}^{m}\sum_{j=1}^{m} \alpha_i \alpha_j y_i y_j x_i \cdot x_j \\ \text{subject to } \begin{cases} \sum_{i=1}^{m} \alpha_i y_i = 0 \\ \alpha_i \geq 0 \end{cases} \end{cases} \tag{2.18}$$

The optimum Lagrange multipliers are next determined by setting the gradient of Q to zero and solving the resulting system.

Then, the optimum vector w can be computed from (2.12) [Haykin, 1999]:

$$w = \sum_{i=1}^{m} \alpha_i y_i x_i$$

As b is concerned, it can be obtained from any of the equalities of (2.14), when $\alpha_i \neq 0$. Then:

$$y_i(w \cdot x_i - b) - 1 = 0 \Rightarrow$$

$$y_i(\sum_{j=1}^{m} \alpha_j y_j x_j \cdot x_i - b) = 1 \Rightarrow$$

$$\sum_{j=1}^{m} \alpha_j y_j x_j \cdot x_i - b = y_i \Rightarrow$$

$$b = \sum_{j=1}^{m} \alpha_j y_j x_j \cdot x_i - y_i$$

Note that we have equalled $1/y_i$ to y_i above, since y_i can be either 1 or -1.

Although the value for b can be thus directly derived from only one such equality when $\alpha_i \neq 0$, it is nevertheless safer to compute all the b values and take their mean as the final result.

In the reached solution to the constrained optimization problem, those points for which $\alpha_i > 0$ are the support vectors and they can also be obtained as the output of the SVM.

Finally, the class for a test sample x' is predicted based on the sign of the decision function with the found coefficients w and b applied to x' and the inequalities in (2.4):

$$class(x') = sgn(w \cdot x' - b)$$

2.3.3 Linearly Nonseparable Data

Since real-world data are not linearly separable, it is obvious that a linear separating hyperplane is not able to build a partition without any errors. However, a linear separation that minimizes training error can be tried as a solution to the classification problem [Haykin, 1999].

The separability statement can be relaxed by introducing slack variables $\xi_i \geq 0$ into its formulation [Cortes and Vapnik, 1995]. This can be achieved by observing the deviations of data samples from the corresponding supporting hyperplanes, which designate the ideal condition of data separability. These variables may then indicate different nuanced digressions (Fig. 2.3), but only a $\xi_i > 1$ signifies an error of classification.

Minimization of training error is achieved by adding the indicator of an error (slack variable) for every training data sample into the separability statement and, at the same time, by minimizing their sum.

For every sample $i = 1, 2, ..., m$, the constraints in (2.5) subsequently become:

$$y_i(w \cdot x_i - b) \geq 1 - \xi_i, \tag{2.19}$$

where $\xi_i \geq 0$.

Simultaneously with (2.19), the sum of misclassifications must be minimized:

$$\text{minimize}\, C \sum_{i=1}^{m} \xi_i. \tag{2.20}$$

$C > 0$ is a parameter of the methodology and is employed for the penalization of errors.

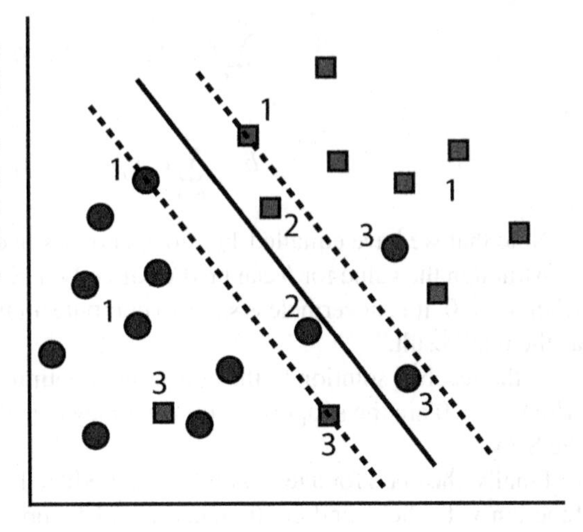

Fig. 2.3 Different data placements in relation to the separating and supporting hyperplanes. Corresponding indicators of errors are labeled by 1, 2 and 3: correct placement, $\xi_i = 0$ (label 1), margin position, $\xi_i < 1$ (label 2) and classification error, $\xi_i > 1$ (label 3).

Therefore, the optimization problem changes to (2.21):

$$\begin{cases} \text{find } w \text{ and } b \text{ as to minimize } \frac{\|w\|^2}{2} + C\sum_{i=1}^{m} \xi_i, C > 0 \\ \text{subject to } y_i(w \cdot x_i - b) \geq 1 - \xi_i, \xi_i \geq 0, \text{ for all } i = 1,2,...,m \end{cases} \tag{2.21}$$

This formulation still obeys the SRM principle as the VC-dimension is once more minimized and separation of training data with as few exceptions as possible is again achieved, both through (2.19) and (2.20).

From the formulation in (2.11), the Lagrangian function changes in the following way [Burges, 1998], where variables α_i and μ_i, $i = 1,2,...,m$, are the Lagrange multipliers:

$$L(w,b,\xi,\alpha,\mu) = \frac{1}{2}\|w\|^2 + C\sum_{i=1}^{m} \xi_i - \sum_{i=1}^{m} \alpha_i \left[y_i(w \cdot x_i - b) - 1 + \xi_i\right] -$$

$$\sum_{i=1}^{m} \mu_i \xi_i,$$

The introduction of the μ_i multipliers is related to the inclusion of the ξ_i variables in the relaxed formulation of the PP.

Application of the KKTL conditions to this new constrained optimization problem leads to the following lines: [Burges, 1998]:

$$\frac{\partial L(w,b,\xi,\alpha,\mu)}{\partial w} = w - \sum_{i=1}^{m} \alpha_i y_i x_i = 0 \Rightarrow w = \sum_{i=1}^{m} \alpha_i y_i x_i \tag{2.22}$$

$$\frac{\partial L(w,b,\xi,\alpha,\mu)}{\partial b} = \sum_{i=1}^{m} \alpha_i y_i = 0 \tag{2.23}$$

$$\frac{\partial L(w,b,\xi,\alpha,\mu)}{\partial \xi_i} = C - \alpha_i - \mu_i = 0 \Rightarrow \alpha_i + \mu_i = C \tag{2.24}$$

The KKTL conditions also require that, for every $i = 1,2,...,m$, the subsequent equalities hold:

$$\alpha_i \left[y_i(w \cdot x_i - b) - 1 + \xi_i\right] = 0 \tag{2.25}$$

$$\mu_i \xi_i = 0 \tag{2.26}$$

We additionally refer to the relaxed separability statement and the conditions for positive slack variables ξ_i and Lagrange multipliers α_i and μ_i for every $i = 1,2,...,m$:

$$y_i(w \cdot x_i - b) - 1 + \xi_i \geq 0$$

$$\xi_i \geq 0$$

$$\alpha_i \geq 0$$

$$\mu_i \geq 0$$

After term by term expansion, the Lagrangian function is then transformed to:

$$L(w,b,\xi,\alpha,\mu) = \sum_{i=1}^{m} \alpha_i - \frac{1}{2}\sum_{i=1}^{m}\sum_{i=1}^{m} \alpha_i\alpha_j y_i y_j x_i \cdot x_j + C\sum_{i=1}^{m} \xi_i - \sum_{i=1}^{m} \alpha_i\xi_i - \sum_{i=1}^{m} \mu_i\xi_i$$

From (2.26), the last term of the Lagrangian becomes zero and following (2.24) and expanding the third term, one obtains:

$$\begin{aligned} L(w,b,\xi,\alpha,\mu) &= \sum_{i=1}^{m} \alpha_i - \frac{1}{2}\sum_{i=1}^{m}\sum_{i=1}^{m} \alpha_i\alpha_j y_i y_j x_i \cdot x_j + \sum_{i=1}^{m} (\alpha_i + \mu_i)\xi_i - \sum_{i=1}^{m} \alpha_i\xi_i \\ &= \sum_{i=1}^{m} \alpha_i - \frac{1}{2}\sum_{i=1}^{m}\sum_{i=1}^{m} \alpha_i\alpha_j y_i y_j x_i \cdot x_j + \sum_{i=1}^{m} \alpha_i\xi_i + \sum_{i=1}^{m} \mu_i\xi_i - \sum_{i=1}^{m} \alpha_i\xi_i \\ &= \sum_{i=1}^{m} \alpha_i - \frac{1}{2}\sum_{i=1}^{m}\sum_{i=1}^{m} \alpha_i\alpha_j y_i y_j x_i \cdot x_j \end{aligned}$$

Consequently, the following corresponding DP is obtained:

$$\begin{cases} \text{find } \{\alpha_i\}_{i=1,2,...,m} \text{ as to maximize } Q(\alpha) = \sum_{i=1}^{m} \alpha_i - \frac{1}{2} \sum_{i=1}^{m} \sum_{j=1}^{m} \alpha_i \alpha_j y_i y_j x_i \cdot x_j \\ \text{subject to } \begin{cases} \sum_{i=1}^{m} \alpha_i y_i = 0 \\ 0 \leq \alpha_i \leq C \end{cases}, C > 0 \end{cases} \tag{2.27}$$

The second constraint is obtained from (2.24) and the condition that $\mu_i \geq 0$, for every sample $i = 1, 2, ..., m$.

The optimum value for w is again computed as:

$$w = \sum_{i=1}^{m} \alpha_i y_i x_i$$

Coefficient b of the hyperplane can be determined as follows [Haykin, 1999]. If the values α_i obeying the condition $\alpha_i < C$ are considered, then from (2.24) it results that for those i $\mu_i \neq 0$. Subsequently, from (2.26) we derive that $\xi_i = 0$, for those certain i. Under these circumstances, from (2.25) and (2.22), one obtains the same formulation as in the separable case:

$$y_i(w \cdot x_i - b) - 1 = 0 \Rightarrow b = \sum_{j=1}^{m} \alpha_j y_j x_j \cdot x_i - y_i.$$

It is again better to take b as the mean value resulting from all such equalities. Those points that have $0 < \alpha_i < C$ are the support vectors.

2.4 Support Vector Machines with Nonlinear Learning

If a linear hyperplane is not able to provide satisfactory results for the classification task, then is it possible that a nonlinear decision surface can do the separation? The answer is affirmative and is based on the following result.

Theorem 2.1. *[Cover, 1965] A complex pattern classification problem cast in a high-dimensional space nonlinearly is more likely to be linearly separable than in a low-dimensional space.*

The above theorem states that an input space can be mapped into a new feature space where it is highly probable that data are linearly separable provided that:

1. The transformation is nonlinear.
2. The dimensionality of the feature space is high enough.

The initial space of training data samples can thus be nonlinearly mapped into a higher dimensional feature space, where a linear decision hyperplane can be subsequently built. The decision hyperplane achieves an accurate separation in the feature space which corresponds to a nonlinear decision function in the initial space (see Fig. 2.4).

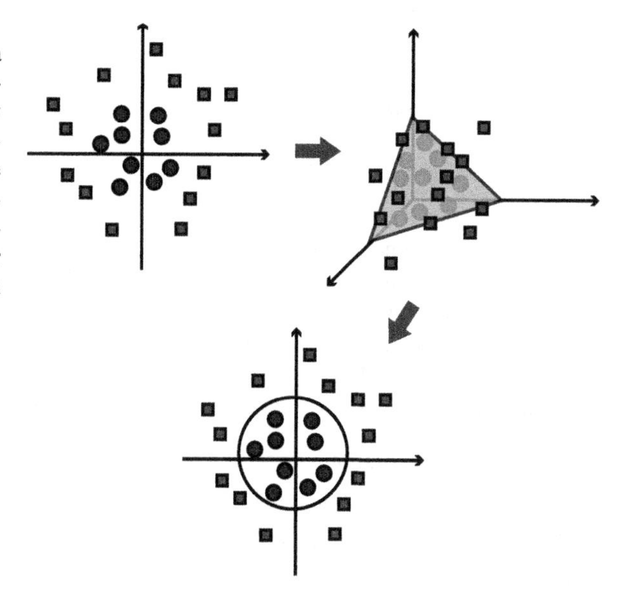

Fig. 2.4 The initial data space with squares and circles (up left) is nonlinearly mapped into the higher dimensional space, where the objects are linearly separable (up right). This corresponds to a nonlinear surface discriminating in the initial space (down).

The procedure therefore leads to the creation of a linear separating hyperplane that minimizes training error as before, but this time performs in the feature space. Accordingly, a nonlinear map $\Phi : \mathbb{R}^n \rightarrow H$ is considered and data samples from the initial space are mapped by Φ into H.

In the standard solving of the SVM optimization problem, vectors appear only as part of scalar products; the issue can be thus further simplified by substituting the dot product by a kernel, which is a function with the property that [Courant and Hilbert, 1970]:

$$K(x,y) = \Phi(x) \cdot \Phi(y), \tag{2.28}$$

where $x, y \in \mathbb{R}^n$.

SVMs require that the kernel is a positive (semi-)definite function in order for the standard solving approach to find a solution to the optimization problem [Boser et al, 1992]. Such a kernel is one that satisfies Mercer's theorem from functional analysis and is therefore required to be a dot product in some space [Burges, 1998].

Theorem 2.2. *[Mercer, 1908]*

Let K(x,y) be a continuous symmetric kernel that is defined in the closed interval $a \leq x \leq b$ *and likewise for y. The kernel K(x,y) can be expanded in the series*

$$K(x,y) = \sum_{i=1}^{\infty} \lambda_i \Phi(x)_i \Phi(y)_i$$

with positive coefficients, $\lambda_i > 0$ for all i. For this expansion to be valid and for it to converge absolutely and uniformly, it is necessary that the condition

$$\int_a^b \int_a^b K(x,y)\psi(x)\psi(y)dxdy \geq 0$$

holds for all $\psi(\cdot)$ for which

$$\int_a^b \psi^2(x)dx < \infty$$

Restricting the kernel to be positive (semi-)definite has two drawbacks [Mierswa, 2006b]. On the one hand, it is difficult to check Mercer's condition for a newly constructed kernel. On the other hand, kernels that fail to meet the conditions of the theorem might have proven to achieve a better separation of the training samples.

When applying SVMs for a classification task, there are a couple of classical kernels that had been demonstrated to meet Mercer's condition [Vapnik, 1995b]:

- the polynomial kernel of degree p: $K(x,y) = (x \cdot y)^p$
- the radial basis function kernel: $K(x,y) = e^{-\sigma\|x-y\|^2}$,

where p and σ are parameters of the SVM.

One may state the DP in this new case by simply replacing the dot product between data points with the chosen kernel, as below:

$$\begin{cases} \text{find } \{\alpha_i\}_{i=1,2,\ldots,m} \text{ as to maximize } Q(\alpha) = \sum_{i=1}^{m} \alpha_i - \frac{1}{2}\sum_{i=1}^{m}\sum_{j=1}^{m} \alpha_i\alpha_j y_i y_j K(x_i,x_j) \\ \text{subject to } \begin{cases} \sum_{i=1}^{m} \alpha_i y_i = 0 \\ 0 \leq \alpha_i \leq C \end{cases}, C > 0 \end{cases} \tag{2.29}$$

As generally one is not able to construct the mapping Φ from the kernel K, the value for the optimum vector w cannot always be determined explicitly from:

$$w = \sum_{i=1}^{m} \alpha_i y_i \Phi(x_i)$$

Consequently, one usually has to directly determine the class for a new data sample x', as follows:

$$class(x') = sgn(w \cdot \Phi(x') - b)$$

Therefore, by replacing w with $\sum_{i=1}^{m} \alpha_i y_i \Phi(x_i)$, one gets:

$$\begin{aligned} class(x') &= sgn(w \cdot \Phi(x') - b) \\ &= sgn(\sum_{i=1}^{m} \alpha_i y_i \Phi(x) \cdot \Phi(x_i) - b) \\ &= sgn(\sum_{i=1}^{m} \alpha_i y_i K(x,x_i) - b) \end{aligned}$$

One is left to determine the value of b. This is done by replacing the dot product by the kernel in the formula for the linear case, *i.e.* when $0 < \alpha_i < C$:

$$b = \sum_{j=1}^{m} \alpha_j y_j K(x_j,x_i) - y_i,$$

and taking the mean of all the values obtained for b.

2.5 Support Vector Machines for Multi-class Learning

Multi-class SVMs build several two-class classifiers that separately solve the corresponding tasks. The translation from multi-class to two-class is performed through different systems, among which one-against-all, one-against-one or decision directed acyclic graph are the most commonly employed.

Resulting SVM decision functions are considered as a whole and the class for each sample in the test set is decided by the corresponding system [Hsu and Lin, 2004] .

2.5.1 *One-Against-All*

The one-against-all technique [Hsu and Lin, 2004] builds k classifiers. Every i^{th} SVM considers all training samples labeled with i as positive and all the remaining ones as negative.

The aim of every i^{th} SVM is thus to determine the optimal coefficients w and b of the decision hyperplane to separate the samples with outcome i from all the other samples in the training set, such that (2.30) :

$$\begin{cases} \text{find } w^i \text{ and } b^i \text{ as to minimize } \dfrac{\|w^i\|^2}{2} + C \sum_{j=1}^{m} \xi_j^i \\ \text{subject to } y_j(w^i \cdot x_j - b^i) \geq 1 - \xi_j^i, \xi_j^i \geq 0, \text{ for all } j = 1,2,...,m. \end{cases} \tag{2.30}$$

Once the all hyperplanes are determined following the classical SVM solving as in the earlier pages, the class for a test sample x' is given by the category that has the maximum value for the learning function, as in (2.31):

$$class(x') = argmax_{i=1,2,...,k}(w^i \cdot \Phi(x')) - b^i) \tag{2.31}$$

2.5.2 *One-Against-One and Decision Directed Acyclic Graph*

The one-against-one technique [Hsu and Lin, 2004] builds $\frac{k(k-1)}{2}$ SVMs. Every i^{th} machine is trained on data from every two classes, i and j, where samples labelled with i are considered positive while those in class j are taken as negative.

The aim of every SVM is hence to determine the optimal coefficients of the decision hyperplane to discriminate the samples with outcome i from the samples with outcome j, such that (2.32) :

$$\begin{cases} \text{find } w^{ij} \text{ and } b^{ij} \text{ as to minimize } \dfrac{\|w^{ij}\|^2}{2} + C\sum_{l=1}^{m}\xi_l^{ij}, \\ \text{subject to } y_l(w^{ij} \cdot x_l - b^{ij}) \geq 1 - \xi_l^{ij}, \xi_l^{ij} \geq 0, \text{ for all } l = 1,2,...,m \end{cases} \tag{2.32}$$

When the hyperplanes of the $\frac{k(k-1)}{2}$ SVMs are found, a voting method is used to determine the class for a test sample x'. For every SVM, the class of x' is computed by following the sign of its resulting decision function applied to x'. Subsequently, if the sign says x' is in class i, the vote for the i-th class is incremented by one; conversely, the vote for class j is increased by unity. Finally, x' is taken to belong to the class with the largest vote. In case two classes have an identical number of votes, the one with the smaller index is selected.

Classification within the decision directed acyclic graph technique [Platt et al, 2000] is done in an identical manner to that of one-against-one.

For the second part, after the hyperplanes of the $\frac{k(k-1)}{2}$ SVMs are discovered, the following graph system is used to determine the class for a test sample x (Fig. 2.5). Each node of the graph has an attached list of classes and considers the first and last elements of the list. The list that corresponds to the root node contains all k classes. When a test instance x is evaluated, one descends from node to node, in other words, eliminates one class from each corresponding list, until the leaves are reached.

The mechanism starts at the root node which considers the first and last classes. At each node, i vs j, we refer to the SVM that was trained on data from classes i and j. The class of x is computed by following the sign of the corresponding decision function applied to x. Subsequently, if the sign says x is in class i, the node is exited via the right edge; conversely, we exit through the left edge. We thus eliminate the wrong class from the list and proceed via the corresponding edge to test the first and last classes of the new list and node. The class is given by the leaf that x eventually reaches.

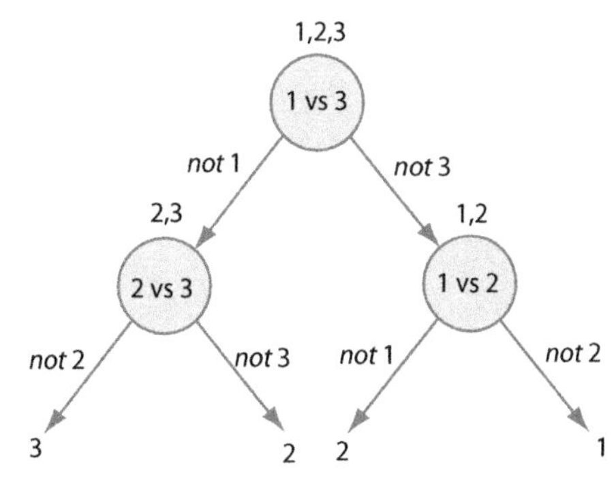

Fig. 2.5 An example of a 3-class problem labeled by a decision directed acyclic graph

2.6 Concluding Remarks

SVMs provide a very interesting and efficient vision upon classification. They pursue a geometrical interpretation of the relationship between samples and decision surfaces and thus manage to formulate a simple and natural optimization task.

On the practical side, when applying the technique for the problem at hand, one should first try a linear SVM (with possibly some errors) and only after this fails, turn to a nonlinear model; there, a radial kernel should generally do the trick.

Although very effective (as demonstrated by their many applications, like those described in [Kramer and Hein, 2009], [Kandaswamy et al, 2010], [Li et al, 2010], [Palmieri et al, 2013], to give only a few examples of their diversity), the standard solving of the reached optimization problem within SVMs is both intricate, as seen in this chapter, and constrained: the possibilities are limited to the kernels that obey Mercer's theorem. Thus, nonstandard possibly better performing decision functions are left aside. However, as a substitute for the original solving, direct search techniques (like the EAs) do not depend on the condition whether the kernel is positive (semi-) definite or not.

Part II
Evolutionary Algorithms

The second part of this book presents the essential aspects of EAs, mostly those related to the application of this friendly paradigm to the problem at hand. This is not a thorough description of the field, it merely emphasizes the must-have knowledge to understand the various EA approaches to classification.

Chapter 3
Overview of Evolutionary Algorithms

It is not the strongest or the most intelligent who will survive but those who can best manage change.
Charles Darwin

3.1 Goals of This Chapter

Every aspect of our daily life implies a search for the best possible action and the optimal choice for that act. Most standard optimization methods require the fulfilment of certain constraints, imply convergence issues or use single point movement. Among them, EAs represent a flexible and adaptable alternative, with classes of methods based on principles of evolution and heredity, with populations of potential solutions and only some basic knowledge of mathematics.

This chapter explains the concepts revolving around the field of evolutionary computation (EC) (Sect. 3.2), presents a general scheme of an EA and describes its main components (Sect. 3.3). The choices for representation and evolutionary operators are only summarized; more emphasis is given to the types employed further on within the approaches that are put forward by this book. Finally, Sect. 3.11 outlines the traditional EA approaches to classification as a standard reference for the envisaged contributions.

3.2 The Wheels of Artificial Evolution

Evolution is the driving force behind all life forms in nature, since, even under very cruel conditions, diverse species that make up our world manage to survive and adapt in their own niches. As nature represents a great source of inspiration for scientists, the development of an artificial optimization framework to mimic evolution and heredity and transform their laws into arithmetical operators came natural and it materialized through the field of EAs.

When solving a problem in the EC context, an environment is created and filled with a population of individuals. These represent a collection of candidate solutions for the considered task. Learning is viewed as a process of continuous adaptation of the individuals to the initially unknown environment. This acclimatization is achieved through reproduction, recombination and mutation. The fitness of the

C. Stoean and R. Stoean, *Support Vector Machines and Evolutionary Algorithms for Classification*, Intelligent Systems Reference Library 69,
DOI: 10.1007/978-3-319-06941-8_3, © Springer International Publishing Switzerland 2014

individuals is closely related to how well they adjust to the environment and represents their chance of survival and multiplication.

Provided that the environment can only host a limited number of individuals and given their capacity to reproduce, selection is inevitable if population size is forbidden to grow exponentially. Obviously, natural selection favors those individuals that best adapt to the environmental settings – historically called by Darwin as *survival of the fittest* – hence the best ones survive and reproduce and evolution progresses step by step. Occasional mutations take place in order to introduce new testing material. Thus, the constitution of the population changes as time passes and it evolves, offering in the end the most adequate solution(s) for the considered problem.

As a correspondence between natural evolution and problem solving, the environment is said to represent the problem, individuals are candidate solutions and the fitness corresponds to the quality of the solution. The quality of a candidate determines the chance of the considered individual to be used as a seed for building new potential solutions. As in nature, mutation also sporadically attacks certain traits of an individual. The next generation is presumably a better one than that of its parents.

There are many approaches that simulate evolution: genetic algorithms (GAs), evolution strategies, genetic programming, evolutionary programming [Fogel, 1995], [Bäck, 1996], [Michalewicz, 1996], [Bäck et al, 1997], [Dumitrescu et al, 2000], [Sarker et al, 2002], [Eiben and Smith, 2003], [Schwefel et al, 2003]. All of them imply the use of selection of individuals in a population, reproduction, random variation and competition, which are the essences of evolution, both in nature or inside a computer [Fogel, 1997]. EAs are simple, general and fast, with several potential solutions that exist at the same time. They are semi-probabilistic techniques that combine local search - the exploitation of the best available solutions at some point - with global search - the exploration of the search space. The different properties of continuity, convexity or derivability that have been standardly required in classical optimization for an objective function are of no further concern within EAs.

Real-world tasks have immense solution search spaces and the high number of local optima hardens the process of finding the global optimum in reasonable time. Moreover, the problems can have dynamic components that can change the location of the optima in time and, therefore, the technique that is considered for solving them must adapt to the changes. Additionally, the final solution may have nonlinear constraints that have to be fulfilled (constrained problems) or may have objectives that are in conflict (multiobjective problems). EAs represent an appropriate alternative for solving such problems. They are population-based approaches, thus multiple regions of the search space can be simultaneously explored. This is especially advantageous when dealing with a multimodal search space where an EA keeps track of several optima in parallel and maintains diversity in the population of solutions, with the aim of performing a better exploration of the search space for finding the global optimum. When the considered problem is dynamic, the solutions in the population continuously adapt to the changing landscape and move towards the new optima. For multiobjective problems [Coello et al, 2007], [Branke et al, 2008], EAs provide, in the end of the evolution process, a set of trade-off solutions for the conflicting objectives, while traditional techniques only produce one solution at the end of a run. For constrained problems, EAs offer a set of feasible and unfeasible solutions. Probably their main advantage and the reason why they are frequently used nowadays is that they can be applied to any type of optimization problem, be that it is continuous or discrete. Another important advantage is that they can be easily hybridized with existing techniques.

It may be tempting to perceive EAs as the applicable solver for any optimization problem, which is false.

> If there is already a traditional method that solves a given problem, EAs should not be used [Schwefel, 1997].

At least one should not expect their alternative solving to be both better or less computationally expensive. The computational effort indeed represents an important drawback of EAs as many candidate solutions have to be evaluated in the evolutionary process. But even if EAs alone do not necessarily provide the best possible solution for the problem at hand, they can be used for adding further improvements to solutions obtained by other means. Artificial evolution thus represents an optimization process that does not reach perfection, but still can obtain highly precise solutions to a large scale of optimization problems. Conversely, an EC technique is better than a random search strategy.

There is a continuous interest for researchers in this field as well as from completely different areas towards the application of EAs for solving practical optimization problems. They are also called *the Swiss army knife* of metaheuristics and they owe their charm to their simplicity and flexibility.

3.3 What's What in Evolutionary Algorithms

Artificial evolution performs a precise algorithmic cycle which closely follows its natural counterpart. Figure 3.1 intuitively shows a typical evolutionary flow. Given a population of individuals, the environmental pressure causes natural selection, the survival of the fittest, and consequently the average fitness along the generations gradually increases. Supposing the fitness function has to be maximized, a set of randomly generated candidate solutions are created in the domain of the function and are evaluated – the better individuals are considered those with higher values for the fitness function. Based on the computed fitness values, a part of the individuals are selected to be the parents of a new generation of individuals. Descendants are obtained through variation, i.e., by applying recombination and/or mutation to the previously chosen individuals. Recombination takes place between two or more individuals and one or more descendants (or offspring) are obtained; descendants borrow particularities from each of the parents. When mutation is applied to a potential solution, the result is one new individual that is usually only slightly different from its parent. After applying the variation operators, a set of new individuals is obtained that will fight for survival with the old ones for a place in the next generation. The candidate solutions that are fitter are again advantaged in this competition. The evolutionary process – parent selection - variation - survival selection – resumes and usually stops after a predefined computational limit is reached.

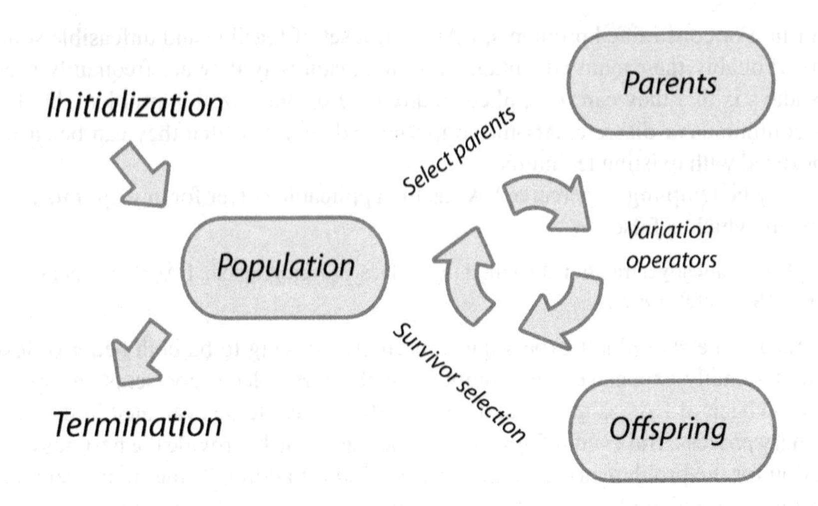

Fig. 3.1 The cycle of an EA

We have introduced several EA-related concepts in the previous paragraph. They are summarized below:

1. Initialization
2. Representation of candidate solutions
3. Population model
4. Fitness function
5. Selection
   - Parent selection strategy
   - Survival selection strategy
6. Variation operators
7. Termination criterion

An EA for a specific problem is completely built only when each of these components is carefully specified.

A general formulation of a canonical EA is outlined by Algorithm 3.1.

Although initialization and a termination criterion are also common for the traditional approaches to optimization, the presence of selection and variation operators needs further argumentation apart from that of perfectly mimicking nature. Variation operators have the role of introducing new candidates into the population and, in this way, explore the search space. But by applying them alone, worse solutions might be encountered; the task of avoiding the decrease in quality of the populations with respect to the fitness evaluations is achieved by selection. Selection exploits the fitter candidate solutions, so the average value of the population fitness consequently increases. In conclusion, variation operators create diversity in the population, having an explorative role, while selection favors the better individuals, having therefore an exploitative task.

Algorithm 3.1 A standard EA.

Require: An optimization problem
Ensure: The best obtained individual(s)
  begin
  Initialization;
  Evaluation;
  while termination condition is not satisfied **do**
    Selection for reproduction;
    Recombination;
    Mutation;
    Evaluation;
    Survivor selection for the next generation;
  end while
  return best obtained solution(s)
  end

In order to reach the optimum of the problem to be solved, a good equilibrium between exploration and exploitation has to be established [Eiben and Smith, 2003]. When performing too much exploitation and only little exploration, some promising regions from the search space might remain unexplored, so the best solutions might not be found at all. In such a situation, there is a high probability that the search process remains blocked into a local optimum. Otherwise, if there is too much exploration and only little exploitation, the search process is significantly slowed down and the time needed for the convergence of the algorithm to the optimum might become too large.

It must also be underlined that EAs are stochastic optimizers. When selection is applied, the fitter individuals have higher chances to be chosen than the less fit ones. However, even the weak individuals, with respect to the fitness evaluation, have a (smaller) chance to be selected. In the same way, when recombination is applied to two or more candidates, the offspring genes are randomly chosen from each of the parents. In the case of mutation, the parts of the individual that are changed are also picked in a random fashion.

In the following sections, we will present the evolutionary components in detail. The most common choices are briefly discussed and those used in further experiments throughout this book are emphasized.

3.4 Representation

The first thing to do when solving a problem through EAs is to set up a bridge between the problem space and the EA space. The possible solutions of the original problem (phenotypes) have to be encoded into individuals within the EA (genotypes).

While sometimes the phenotypic and the genotypic spaces may coincide, other times they can be completely different. For instance, if the optimization problem is in an integer domain, a real-valued genotype representation can be chosen, so the two spaces would be the same, or it could be decided for a binary representation. A solution of the form "21" from the phenotypic space would then be represented as "10101" in the genotypic space. The way the individual representation is chosen depends very much on the problem to be solved. In conclusion, a mapping is therefore necessary to transfer the solution of the problem into the EA space (encoding) and an inverse mapping will be also needed, in order to transform the EA result back into the problem solution (decoding).

Genotypes or individuals are also called chromosomes, but throughout the book we will mostly refer to them as individuals. The individual is composed of genes. A gene is located at a particular position in the chromosome. A gene may contain several values or may have several forms. Each value of a gene is referred to as an allele of that gene.

A binary representation is used for problems searching for 0/1 values (such as truth values) for certain variables. It is however often inappropriately chosen for solving other tasks, just because it is typical of the widely used EA subclass of GAs.

An integer encoding is most appropriate when solutions have to be represented in a discrete space. An example would be the search for optimal values for a set of variables that are instantiated with integers. Permutation problems where one must determine an optimal permutation of elements that lead to a potential solution can also be represented by integer values.

When the values represented by genes come from a continuous space, a real-valued encoding is employed. Every individual is then a vector of real components and each of them takes values in a certain domain and must be kept in that interval all throughout evolution.

The chosen representation is strongly related to the type of recombination and mutation operators that are used. Since the techniques we discuss in this book use only binary and real-valued encodings, we will summarize only some variation operators that correspond to these representations.

3.5 The Population Model

The role of the population is that of preserving all candidate solutions at one time; it consists of a set of genotypes that are not necessarily all different from each other. The population as a unit is the one that evolves, not the individuals. When selection is applied for choosing the parents of the population that will form the next generation, it picks the fitter candidates from the current population. Also, the resulting offspring are intended to replace the former individuals in a fitter future population.

The population size represents the number of individuals that the population contains. Usually, the population size, which is a parameter of the EA, is constant from the start to the end of the algorithm. However, an algorithm that contains a population with decreasing size is presented in Chap. 4.

The initialization of the population implies addressing population size together with representation. Each gene of every individual generally takes a random value from its domain of representation. Sometimes, the EA may start using as an initial population a fixed set of candidate solutions obtained by other methods.

3.6 Fitness Evaluation

The role of the fitness function (or evaluation function) is to measure the extent by which individuals adapted to the environment. It is mostly used by selection and thereby makes improvements possible.

The fitness function is defined on the genotypic space and its values are usually real numbers, in order to be able to make comparisons between the qualities of different individuals. Returning to the previous example of a binary representation, and presuming that a real-valued function is to be optimized, e.g. $f(x) = x^2$, the fitness of the genotype 10101 is $21^2 = 441$.

In many cases, the objective function, which is the name used in the original context of the problem, coincides with the fitness function or the fitness function is a transformation of the given objective function.

3.7 The Selection Operator

Selection appears twice during an evolutionary cycle. First, there is selection for reproduction, when parents of the next generation are chosen (parent or mating selection). Secondly, there is selection for replacement, when individuals that will form the next generation are chosen from the offspring and the current population (survivor selection). Both selection types are responsible for quality enhancement.

3.7.1 Selection for Reproduction

The role of parent selection is that of choosing which of the individuals in the current population should be considered to undergo variation in order to create offspring, based on their quality. Parent selection is typically probabilistic: high-quality individuals have a good chance to be selected for reproduction, while low-quality ones have small chances of becoming parents.

The selection operator does not create new candidate solutions. It is solely responsible for selecting relatively good solutions from the population and discarding the remaining candidates. As the population size usually remains constant, the selection conducts to the placement of multiple copies of certain individuals in a new population by removing inferior solutions.

The basic idea is that individuals with a better fitness must have a higher probability of being selected. Nevertheless, selection operators differ in the way the chances are assigned to better solutions. Some operators sort the individuals in the population according to their fitness and then deterministically choose some few best individuals. Meanwhile, other operators assign a selection probability to each individual which is dependent on its fitness. In this case, there exists the possibility of selecting a bad solution and, at the same time, of rejecting a good one. However, this can be an advantage: the fittest individuals in a population can be connected to a suboptimal region in the fitness landscape and, by using a deterministic selection, the EA would evidently converge to the wrong, suboptimal solution. If, conversely, a probabilistic selection is employed here, diversity is maintained for a higher number of generations by selecting some less fit individuals. Therefore, more exploration is performed and the EA would eventually be prevented from converging to a wrong solution.

Two of the most popular schemes, i.e. the proportional and the tournament selections, are briefly mentioned in the following paragraphs.

Proportional selection implies that the number of copies an individual will have is directly proportional to its fitness. A solution having double the fitness of another solution will also have twice as many copies in the selected population. The most commonly used form of implementing selection probabilities within the proportional type is the roulette-wheel (or Monte-Carlo) mechanism, where each individual in the population occupies a section on a roulette that has a size directly proportional to its fitness. Then, the wheel is spun as many times as the population size and at every turn the solution indicated by the wheel is selected. Evidently, solutions with better fitness have higher chances to be selected (and therefore to have several copies in the population) as their sections on the wheel are proportional to their fitness.

Algorithm 3.2 puts forward the basic steps for implementing the proportional scheme. A probability selection is computed for each individual like in (3.1), by referring the sum of the fitness evaluations for all individuals in the population. Then, the sections of the roulette-wheel are computed as in (3.2). Next, the wheel is turned n times by randomly generating a number in the [0, 1] interval and determining its correspondingly chosen section: that is the one pointing to the individual that is selected.

$$P_{sel}(x_i) = \frac{f(x_i)}{\sum_{j=1}^{n} f(x_j)} \tag{3.1}$$

$$a_i = \sum_{j=1}^{i} P_{sel}(x_j) \tag{3.2}$$

There are however several limitations of the proportional selection scheme. First of all, if there exists a very fit individual in comparison to all the others in the population, proportional selection selects a very high number of copies of that individual and this leads to a loss of diversity in the population which conducts to premature convergence. On the other hand, if all candidates have very similar evaluations, which usually happens later on in the evolutionary process, the roulette-wheel will be marked approximately equally for all individuals in the population and all of

Algorithm 3.2 Proportional selection.

Require: The population that consists of n individuals
Ensure: n individuals selected for reproduction
begin
for $i = 1$ to n **do**
Compute the probability $P_{sel}(x_i)$ to select individual x_i as in (3.1);
end for
for $i = 1$ to n **do**
Compute the roulette section a_i for individual x_i as in (3.2);
end for
$i \leftarrow 1$;
while $i \leq n$ **do**
Generate a random number $r \in [0, 1]$;
$j \leftarrow 1$;
while $a_j < r$ **do**
$j \leftarrow j + 1$;
end while
$select_i \leftarrow x_j$;
$i \leftarrow i + 1$;
end while
return the selected individuals
end

them will have almost the same chances of being selected (the effect of random selection). On a different level, it cannot handle negative values for fitness, as they should correspond to sections on the roulette-wheel. Finally, it cannot handle minimization problems directly, but they have to be transformed into a maximization formulation. A way to avoid these last two drawbacks is by using a scaling scheme, where the fitness of every solution is mapped into another interval before marking the roulette wheel [Goldberg, 1989].

These issues of proportional selection are avoided when using the tournament type. For a number of times equal to the population size, one chooses the best solution in a tournament of k individuals. In the simplest form, k is 2, the scheme is called binary tournament selection and the best one of each two solutions is chosen for as many times as to form a population of the same size as before. This selection operator does not depend on whether the fitness values are positive or negative and, when one deals with a minimization problem (instead of a maximization one), the only difference is that the individuals with the smaller fitness value are now selected (instead of the ones with the higher fitness score). The absolute performances of individuals do not count, it is only the actual values they exhibit in relation to one another that are important. Therefore, in this type of selection there is no need for global knowledge on the population as in the proportional scheme.

Due to its simplicity and flexibility, it is the scheme we use the most throughout the book and it is outlined in Algorithm 3.3.

Algorithm 3.3 Tournament selection.

Require: The population that consists of n individuals
Ensure: n individuals selected for reproduction
begin
$i \leftarrow 1$;
while $i \leq n$ **do**
Choose k individuals;
Take the fittest one x from them;
$select_i \leftarrow x$;
$i \leftarrow i + 1$;
end while
return the selected individuals
end

An important parameter and advantage of this selection is given by the tournament size k. For higher values of k, there is a stronger selection pressure, which triggers the choice of above average individuals. Smaller values of k on the contrary also give weaker individuals the chance of being selected, which eventually leads to a better exploration of the search space.

Ranking selection is similar to the proportional scheme, but instead of using the direct fitness values, the individuals are ordered according to their performance and attributed a corresponding rank. It can also treat negative evaluation results and, when minimization is required, the only difference is that ranking has to be inversely performed.

3.7.2 *Selection for Replacement*

Another situation when selection takes place is when it is decided which individuals from the current population and their offspring are retained to form the population of the next generation. In order to preserve the same population size after offspring are obtained via the variation operators, survivor selection has to intervene. This selection decision is usually elitist (the best individuals are preferred) and takes into account the fitness of all individuals (new and old), favoring the fitter candidate solutions.

3.8 Variation: The Recombination Operator

We have seen that the selection operator has the task of focusing search on the most promising regions of the search space. On the other side, the role of the variation operators is to create new candidate solutions from the old ones and increase population diversity. They are representation dependent, as for various encodings, different operators have to be defined [Bäck, 1996], [Bäck et al, 1997], [Eiben and Smith, 2003].

We will thus further discuss the standard operators for introducing variation, i.e., recombination and mutation, along with their most common representation-related choices. Those of particular need for the upcoming approaches of this book are described at larger extent.

Recombination or crossover is a variation operator that is responsible for forming offspring by combining the genes of parent individuals. Recombination represents a stochastic operator, since choices like which parts are to be inherited from one parent and which from the other or the way the two parts are combined depend on a pseudo-random number generator.

When mating (usually two) individuals with different attributes, an offspring (or two) that combine those features is (are) obtained. The aim is to explore the space between the two individuals, in search of a potential solution that has a better quality.

In a more general scenario, recombination can even take place between p individuals ($p > 2$) and p offspring may be obtained [Bäck et al, 1997]; one offspring alone can also be constructed by combining traits from the p parents.

The process usually takes place as follows. For every individual in the current population obtained after the application of the selection for reproduction, a random number in the [0, 1] interval is generated. If it is smaller than the given crossover probability, then the current individual is chosen for recombination. If the number of

---

Algorithm 3.4 The main steps of a recombination process using probability p_r.

Require: A population that consists of n individuals
Ensure: A population obtained after the recombination process

```
begin
i ← 0;
for j = 1 to n do
   Generate q in [0,1]; {select individuals for recombination in vector parent with probability p_r}
   if q < p_r then
      i ← i + 1;
      parent[i] ← individual[j];
   end if
end for
if i is odd then
   Generate q in [0,1]; {pairs of parents should be formed, so either add or remove an individual}
   if q < 0.5 then
      Add a random individual from population to parent;
   else
      Remove a random item from parent;
   end if
end if
Apply the chosen recombination type for each pair of parents;
return the resulting population
end
```

---

chosen individuals is odd, then they recombine as pairs which are randomly formed. Otherwise, one individual is deleted or another one is added from the parents pool, a decision which is also randomly taken. The steps are presented in Algorithm 3.4.

As concerns the different possibilities of recombination, several options can be distinguished, depending on the type of representation – binary, integer or real-valued. The choices are numerous, however only the most common schemes within the binary and real-valued representations are further outlined [Eiben and Smith, 2003], as some of them will be addressed in the algorithms of this book.

For the binary representation, an often used scheme is the one point recombination. A random position is generated which is the point of split for the two parents, as in Fig. 3.2. The resulting offspring take the first part from one parent and the other from the other, respectively. Suppose there are two parent individuals $c = (c_1, c_2, ..., c_m)$ and $d = (d_1, d_2, ..., d_m)$ and we take a random number k from the set $\{1, 2, ..., m\}$. The first offspring copies the first k genes from parent c and the remaining from d, like this: $o_1 = (c_1, c_2, ..., c_k, d_{k+1}, ..., d_m)$; conversely, the second offspring takes the first k genes from parent d and the remaining from c, i.e. $o_2 = (d_1, d_2, ..., d_k, c_{k+1}, ..., c_m)$.

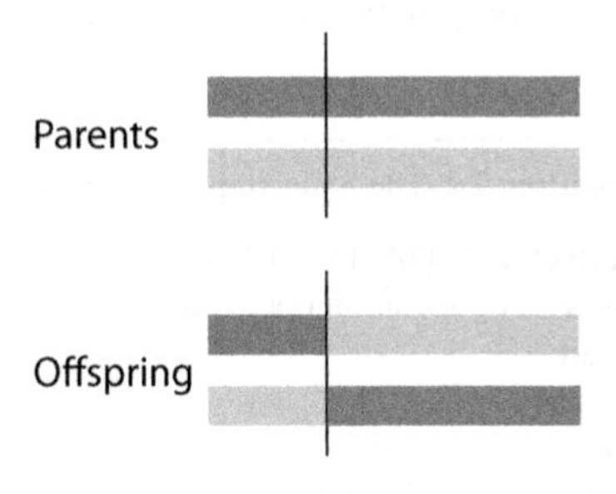

Fig. 3.2 One point crossover for binary representation. The first part of offspring 1 is copied from the first parent and the other section from the second parent, while for the second offspring the complementary parts are inherited from the two parents.

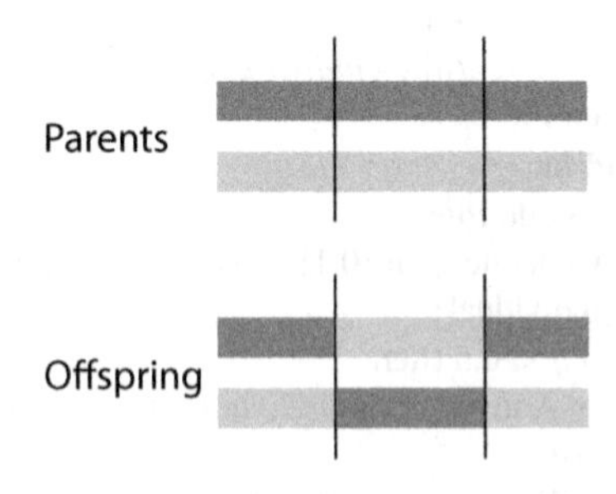

Fig. 3.3 Two point crossover for binary representation. The offspring take alternate sections from the parents. The same type of inheritance occurs when several cut points are considered.

A generalized form is the multiple point recombination, where several cut points are considered: an example of two point crossover can be visualized in Fig. 3.3. Furthermore, within adaptive recombination, the split points also undergo evolution by adapting to previous splits that took place. Additionally, segmented recombination is another variant of the multiple point recombination where the number of points can vary from one individual to another [Eiben and Smith, 2003].

Uniform recombination does not use split points. For every gene of the first offspring it is probabilistically decided which parent gives the value of that component, while the corresponding gene of the second offspring gets the value from the other parent. This operator could also be considered such that the values for both offspring individuals are computed in the same probabilistic manner, but independently.

Shuffle recombination is an add-on to an arbitrary binary recombination scheme and has the advantage of removing positional bias. The genes of the two parents are shuffled randomly, remembering their initial positions. The resulting individuals may then undergo any kind of binary crossover. Resulting offspring are un-shuffled.

Such recombination mechanisms are obviously not suitable for a real-valued encoding. Intermediate recombination presumes that the value for the gene of the offspring is a convex combination of the corresponding values of the parents. We have again two parents $c = (c_1, c_2, ..., c_m)$ and $d = (d_1, d_2, ..., d_m)$ and offspring gene $o_i = \alpha c_i + (1 - \alpha) d_i$, for $\alpha \in [0, 1]$. Parameter α can be randomly chosen at each application of the recombination operator or it can be fixed: it is very often set at 0.5, which leads to a uniform arithmetic recombination. Depending on the number of recombined genes for the offspring, there are three types of intermediate recombination: single (when one position is changed), simple (when we change all values from some point on) and total (when all genes are affected).

3.9 Variation: The Mutation Operator

Mutation is a unary variation operator. It is also a stochastic one, so the genes whose values are considered to be changed are chosen in a probabilistic manner. When applied to an individual, the resulting offspring contains minor modifications as opposed to the initial individual. Through mutation, individuals that cannot be generated using recombination may be introduced into the population, as it makes all the values of a gene available for the search process.

For every individual in the current population and each gene of that individual, a random number in the [0, 1] interval is generated. If the mutation probability is higher than the generated number, then that gene suffers mutation (see Algorithm 3.5).

Sometimes global search is intended in the initial evolutionary phases and local search (fine tuning) is planned towards the final steps of the EA. In this respect, the mutation probability may decrease with the increase in the number of generations. Another distinct situation concerns those cases where the change of a gene lying at the beginning of an individual could make a significant modification to the individual in question, while the same change at the end of the individual would induce a less significant alteration. Then, while the number of generations passes, the probability of mutation of the first genes in every individual may decrease and that of the final ones increase.

Algorithm 3.5 Mutation applied with probability p_m.

Require: A population that consists of n individuals, each containing m genes
Ensure: A population obtained after the mutation process

```
begin
for i = 1 to n do
    for j = 1 to m do
        Generate q in [0,1];
        if q < p_m then
            Apply the chosen mutation operator to the gene j of the current individual i;
        end if
    end for
end for
return the modified population
end
```

Depending on the specific task and the considered representation, there are several forms of the mutation operator as well. Again only the situations with binary and real-valued encodings are further outlined and the most frequent types are described [Eiben and Smith, 2003].

For the binary encoding, a strong mutation (also called bit flip perturbation) presumes that, when a gene undergoes mutation, 1 changes into 0 and 0 into 1. Thus, the value of the gene to be mutated will change through the formula $c_i = 1 - c_i$. An intuitive representation can be observed in Fig. 3.4.

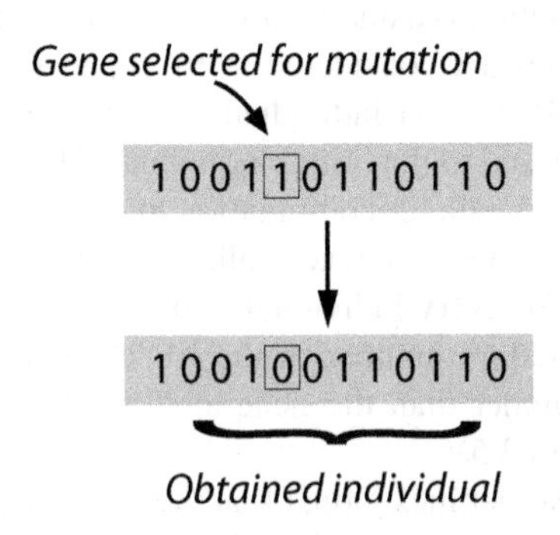

Fig. 3.4 Mutation for binary representation. A position having the value $v \in \{0,1\}$ is randomly chosen in the individual to be mutated and its value is changed to 1 - v.

Weak mutation presumes that the above change does not take place automatically as before, but 1 or 0 is probabilistically chosen and attributed to that position. In this way, the newly generated value could be identical to the old one, so no effective change would actually occur.

For the real-valued individuals, mutation customarily performs a small perturbation in the value of the selected gene. This is induced randomly by a number generated to follow a normal distribution with mean zero and standard deviation given by a parameter called mutation strength. A value of the gene i of an individual c is thus changed according to the formula $c_i = c_i + N(0, ms)$, where ms is the mutation strength parameter.

3.10 Termination Criterion

The stop condition of a typical EA may refer to several criteria like:

- reaching a previously set number of generations,
- not exceeding a predefined number of iterations without achieving any fitness improvement,
- finding a solution with a given accuracy,
- consuming a preset number of fitness evaluation calls,
- allowing the algorithm to run for a certain amount of time,
- population diversity falling below a given threshold,
- a stop button etc.

The solution to the algorithm is the best individual with respect to the fitness function from the last generation, or the best from the entire process, or, sometimes, the entire (or a subset of the) population from the last generation [Bäck, 1996], [Dumitrescu et al, 2000].

3.11 Evolutionary Algorithms for Classification

As the practical side of this book targets classification, the classical evolutionary techniques that had been tailored for this direction will be mentioned. Note that, although there are EA approaches acting both as stand-alone or in hybridization with classification-specific methods, we will present only the former ones. For the latter category, we will specifically outline those related to SVMs in the future Chap. 6 and 7.

The aim of a classification technique may be further conceived as to stepwise learn a set of rules that model the training set as good as possible. When the learning stage is finished, the obtained rules are applied to previously unseen samples within the test set to predict their classes.

EAs may consequently encode IF-THEN rules, while certain mechanisms model their behavior and interaction towards learning. An evolutionary classifier then represents a machine learning system that uses an EA as a rule discovery component [Michalewicz, 1996]. The IF-THEN rules (or productions) are represented through a population that is evolved by an appropriate EA. The rules cover the space of possible inputs and are evolved in order to successfully be applied to the problem to be solved – the fields of application may range from data mining to robotics. On a broader sense, and in connection with the definition of classification from the introductory chapter, an evolutionary classification approach is concerned with the discovery of IF-THEN rules that reproduce the correspondence between given samples and corresponding classes. Given an initial set of training samples, the system learns the patterns, i.e. evolves the classification rules, which are then expected to predict the class of new examples. An IF-THEN rule is imagined as a first-order logic implication where the condition part is made of attributes and the conclusion part is represented by the class.

There are three classical families of evolutionary classifiers [Bacardit and Butz, 2007]:

The Pittsburgh learning strategy [Smith, 1980], [de Jong et al, 1994], [Michalewicz, 1996]: aims to evolve a complete classification rule collection by encoding the entire set into each individual.

The Michigan approach [Holland, 1986], [Wilson, 1995], [Michalewicz, 1996]: considers every individual as the representative of one rule, uses reinforcement learning to reward/penalize the collaboration between rules and achieves the complete optimal rule set as the output of the EA.

Iterative rule learning [Venturini, 1993], [Aguilar-Ruiz et al, 2003], [Bacardit and Butz, 2007]: for each class of the problem, one separate EA run is performed, with all individuals encoding rules for that class and adjusting themselves against the training samples labeled accordingly.

In a Pittsburgh-type evolutionary classifier, each individual represents an entire set of rules. The individuals compete among themselves and only the strong sets survive and reproduce. The Pittsburgh approach uses a typical EA for conducting the learning. What remains to be solved is the representation problem and the way individuals adapt to their environment. Usually, operators from propositional logic, like disjunction and/or conjunction, also appear within the encoding.

In a Michigan-style evolutionary classifier, each individual of the population represents a unique, distinct rule, so the EA evolves a set of rules. Then, the population represents the rule set needed to solve the problem. The goal here is not to obtain the best individual, but to find the best set of individuals (rules) in the end of the algorithm. Usually, representation is divided into two parts – one is the condition part and contains the values for the attributes that appear in the condition of the rule and the other part consists of the conclusion of the rule. A credit assignment system is used in order to reward the better rules or, at the same time, to penalize the worse ones. When new rules are produced through mutation and/or recombination, crowding methods are usually utilized in order to introduce them into the population; in this way, they replace only very similar individuals.

By iterative rule learning, the evolution of a rule for each class is obtained by an equal number of runs of an EA. In every such run, the individuals (rules of one class) are evolved against the training samples of the corresponding class. The fitness expression refers both accuracy and a generality measure that makes the individual cover as much samples as possible.

Finally, note that rules in an IF-THEN format are the traditional means of representing individuals for classification by EAs. Other more complex representations can be employed, like, for instance, a genetic programming approach to rule discovery [Giordana et al, 1994], [Freitas, 1997], [Bojarczuk et al, 2000]. The internal nodes of the individual encode mathematical functions, while the leaf nodes refer the attributes. Given a certain individual, the output of the tree is computed and, if it is greater than a given threshold, a certain outcome of the classification task is predicted.

3.12 Concluding Remarks

We have targeted to show why EAs are easy and straightforward for any task, while their performance as general optimizers is very competitive. Their power and success lie in the simplicity of their mathematical functioning, the naturalness of underlying metaphor and the easy tailoring for any given problem.

After having gone through the introductory chapter into evolutionary computing, the following knowledge has been acquired:

- A general idea of the concepts revolving around EAs.
- The components of a typical EA are subsequently enumerated and are followed by brief descriptions.
- The traditional EA applications for classification are summarized, to set the context for introducing the new evolutionary approaches for this task in Chap. 4, 5, 6 and 7.

Chapter 4
Genetic Chromodynamics

The picture will have charm when each colour is very unlike the one next to it.
Leon Battista Alberti

4.1 Goals of This Chapter

As mentioned in the introductory chapter, for the classification tasks we aim to evolve thresholds for the attributes of the given data training examples. Such vectors of thresholds will be constructed for each class of the problem and we will further refer to them as prototypes. They will efficiently represent the class distributions and therefore well discriminate objects of different labels.

Since EAs eventually converge to a single solution, how can they be useful for a classification problem, where the task is to discover all class prototypes? When the problem landscape has several regions fitter than all the neighboring areas, then the problem is multimodal. Most real-world problems are multimodal and classification certainly also obeys this rule. And EA techniques for multimodal optimization [Eiben and Smith, 2003] can surely achieve prototype determination.

One way of maintaining several prototypes can be done via the most widely used multimodal EA-based methodology, i.e. through search space division: the entire population is separated into subpopulations (or species) that contain only neighboring individuals. And the easiest means of separation is by making use of one (or more) radius (or radii) for creating the subpopulations.

There are many known methods that make use of a radius parameter for separating the entire population into species e.g. [Holland, 1975], [Goldberg and Richardson, 1987], [Li et al, 2002], [Shir and Bäck, 2005], [Cioppa et al, 2007]. Nevertheless, since our personal research interests conducted us towards more deeply studying one recent such methodology, i.e. genetic chromodynamics (GC), it seemed interesting to explore its capabilities towards classification. Through its design, GC is able to concentrate search on many basins of attraction in parallel, so that several optima are found simultaneously; that makes it a good candidate for discovering multiple class prototypes.

The chapter consequently describes the GC framework (Sect. 4.2), outlines a possible enhancement for the paradigm (Sect. 4.3) and presents the adaptation of the algorithm to tackle classification (Sect. 4.4).

C. Stoean and R. Stoean, *Support Vector Machines and Evolutionary Algorithms for Classification*, Intelligent Systems Reference Library 69,
DOI: 10.1007/978-3-319-06941-8_4, © Springer International Publishing Switzerland 2014

4.2 The Genetic Chromodynamics Framework

The framework assumes the following considerations. Subpopulations that are each connected to a local or global optimum of the problem to be solved are built and maintained by introducing a set of restrictions on how selection is applied or the way recombination takes place [Dumitrescu, 2000]. There is not a typical selection scheme considered, as every individual represents a stepping stone for the forming of the new generation, that is each is taken into account for reproduction. If a parent has no similar individuals to itself, then it mutates. For recombination, a local interaction principle is considered, meaning that only individuals similar under a given threshold mate. After either recombination or mutation takes place, the offspring fights for survival with the stepping-stone parent. In order for the subpopulations to become better and better separated with each iteration, a new genetic operator, called merging, is introduced. Merging represents a form of selection that shrinks the population, as some individuals are discarded. The operator "merges" very similar individuals by keeping only the most prolific candidate solutions and it is applied after the perturbation process.

The interplay between the merging and the mating regions strongly impacts convergence properties: a large merging radius leads to the deletion of potential mates of the surviving individuals and thus compromises recombination. Therefore, it is usually chosen to be much smaller than the mating radius.

The complete evolutionary process takes place as follows. First, the initial population is randomly generated. Next, every individual participates in the forming of the new generation. Mating regions around each individual are determined by a radius and, therefore, only neighboring individuals recombine. When no mate is found in the corresponding region of the current individual, the latter produces one offspring by mutation, with a step size that still keeps the descendant in the mating region of its parent. If there is more than one individual near the current, the mate is determined using proportional selection. Then, if the offspring has better fitness than the current individual, it replaces the latter in the population.

Immediately after variation, the merging operator eliminates similarity from the population. For illustration, let an individual c be given. If the distance between c and another individual is very small, i.e., under a given merging radius, then the latter is considered as part of the merging region of c. From the set of all individuals in the merging region of c, only one is kept. Generally, it is the one with the best fitness evaluation that is preserved. Alternatively, other merging schemes may be used (for instance, the mean of the individuals in the merging region).

The unfolding of the GC steps are outlined by Algorithm 4.1.

By reducing the potential partners for recombination of an individual to those lying in its mating region, only individuals that are close to each other recombine, favoring the appearance and maintenance of subpopulations. The offspring replaces the current individual only if fitter. Thus, after a few generations, the candidate solutions will concentrate on the most promising regions of the search space, i.e. those connected to the optima.

Algorithm 4.1 The GC algorithm.

Require: An optimization problem
Ensure: The set of solutions where every one corresponds to an optimum of the problem

```
begin
t ← 0;
Initialize population P(t);
while termination condition is not satisfied do
   Evaluate each individual;
   for all individuals c in the population do
      if mating region of c is empty then
         Apply mutation to c;
         if obtained individual is fitter than c then
            Replace c in P(t+1);
         end if
      else
         Choose one individual from the mating region of c for recombination;
         Obtain and evaluate one offspring;
         if the offspring is fitter than c then
            Replace c in P(t+1);
         end if
      end if
   end for
   repeat
      Consider an individual c as the current one;
      Select all m individuals in the merging region of c, including itself;
      Keep only the fittest individual from the selection in the population of the next gen-
      eration;
   until merging cannot be further applied
   t ← t + 1;
end while
return population P(t) containing the optima
end
```

Merging is an useful operator since it triggers a better computational time by reducing the size of the population, as less fitness evaluations are necessary. Consequently, subpopulations independently evolve and become better separated with each iteration and lead, at convergence, each one to an optimum.

Courtesy of merging, subpopulations become better separated with each iteration and the population size is reduced in a way that preserves the coordinates of each subpopulation. Along with separation, the worse individuals are removed step by step, and the process gradually transforms from a globally–oriented population–based methodology to a parallel local search technique. Therefore, in the end of evolution, it is only one individual that remains connected to every optimum. Each such individual now actually corresponds to a single hill–climber that uses only mutation as a variation operator.

Illustrations of the GC radii-based mating and merging are given in Fig. 4.1 and 4.2.

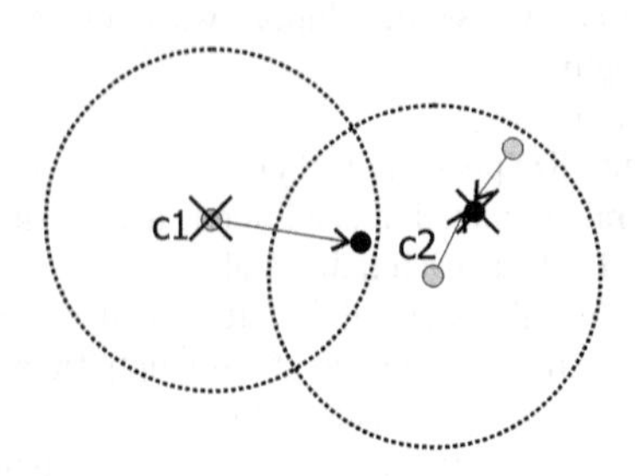

Fig. 4.1 An example of mating within GC. Individual $c1$ is alone in its mating region (dotted circle), so it produces one offspring by mutation. Individual $c2$ selects another parent from its mating region and creates one offspring by recombination that is worse than $c2$ and consequently does not replace its stepping stone parent. Crossed lines indicate individuals with worse fitness that are further replaced.

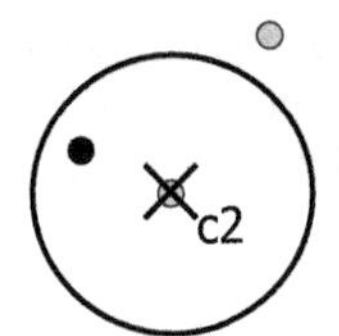

Fig. 4.2 An example of merging within GC that follows the scenario from the mating episode in Fig. 4.1. Individual $c2$ is deleted because there is another individual with better fitness in its merging region (solid circle).

GC can be metaphorically viewed as if all the individuals originally have a distinct color. But, once they mate and merge, the colors of the most promising solutions spread over the entire search space. In the end, there are as many different colors as the number of distinct subpopulations, each connected to an optimum.

The weak point of the methodology is that it makes use of two parameters – the mating and merging radii – which are, for some problems, hard to parameterize. Nevertheless, if the mating parameter is computed using formula (4.2) introduced in [Deb and Goldberg, 1989] and smaller comparable values are tried for the merging radius, one could reach configurations that provide consistent results.

The work [Deb and Goldberg, 1989] proposed a manner of computing the value for the σ_{share} radius leading to the formation of subpopulations in niching EAs, that has become very popular among researchers keen to control such parameters. It uses the radius of the smallest hypersphere containing feasible space, which is given by (4.1):

$$r = \frac{1}{2}\sqrt{\sum_{i=1}^{n}(b_i - a_i)^2} \tag{4.1}$$

where n represents the number of dimensions of the problem at hand and b_i and a_i are the upper and the lower bounds of the i-th dimension. Knowing the number of global optima N_G and being aware that each niche is enclosed by an n-dimensional hypersphere of radius r, the niche radius σ_{share} can be estimated as:

$$\sigma_{share} = \frac{r}{\sqrt[n]{N_G}} \tag{4.2}$$

Although the approximation given by (4.2) is very precise, in many real-world applications, one cannot know the number of global optima in advance and, therefore, in such situations, it cannot be used. For classification, however, we may consider N_G be at least equal to the number of classes of the problem. When the value is higher, more than one prototype is discovered for one class. This situation is convenient as there can be several prototypes that cover well the objects in one class.

4.3 Crowding Genetic Chromodynamics

An enhanced technique within the GC framework that speeds up convergence and, at the same time, looks into the search space for a more accurate approximation of the solutions was proposed in [Stoean et al, 2005]. In contrast to the stepping stone mechanism of the initial algorithm, the first parent is randomly selected. The offspring derived from recombination does not replace any of the parents in particular, but the worst individual (with respect to fitness values) within its replacement radius, a new parameter of the method. The local interaction principle and merging still hold. Mating and merging in the new context of GC with crowding [de Jong, 1975] are depicted in Fig. 4.3 and 4.4 and the complete method is outlined in Algorithm 4.2.

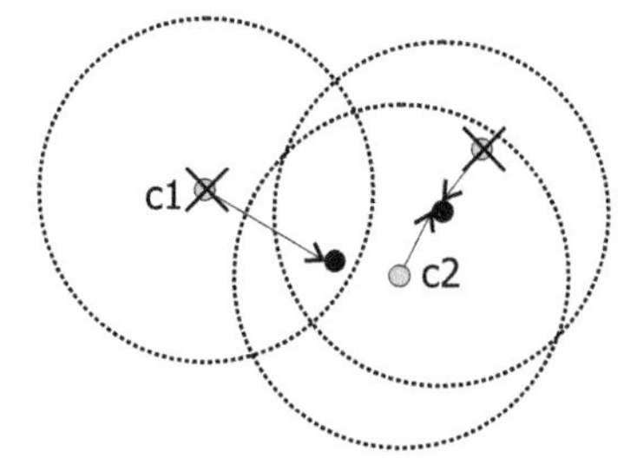

Fig. 4.3 Example of mating within GC with crowding. As in Fig. 4.1, $c1$ and $c2$ each produce one offspring, the former by mutation and the latter through recombination. This time, the offspring of $c2$ replaces its other parent because it is the worst individual in its replacement region and it is also less fit than the offspring.

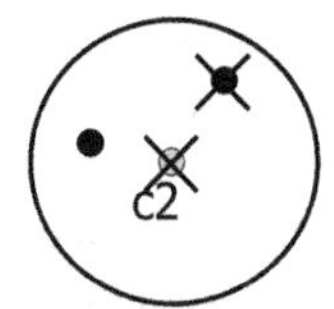

Fig. 4.4 Example of merging within GC with crowding following the scenario from the mating episode in Fig. 4.3. Two individuals are removed, $c2$ and one offspring, because there are now three individuals within merging radius from $c2$.

Algorithm 4.2 GC with crowding.

Require: An optimization problem
Ensure: The set of solutions corresponding each to one local/global optimum of the problem

begin
$t \leftarrow 0$;
Initialize population $P(t)$;
while termination condition is not satisfied **do**
  Evaluate each individual;
  for $i = 1$ to n **do**
    Randomly choose an individual c;
    if mating region of c is empty **then**
      Apply mutation to c;
      if obtained individual is fitter than c **then**
        Replace c in $P(t+1)$;
      end if
    else
      Choose one individual from the mating region of c for recombination;
      Obtain and evaluate one offspring d;
      Find worst individual w within replacement radius of d;
      if d has better fitness than w **then**
        Replace w in $P(t+1)$;
      end if
    end if
  end for
  Apply merging to all individuals;
  $t \leftarrow t + 1$;
end while
return population $P(t)$ containing the optima
end

The aims of the new method were those of preserving the ability of GC to properly locate several or all optima and additionally of speeding up this process. The new model achieves the first goal and the second one is accomplished only for more complex tasks (for instance, for higher dimensional problems).

The original approach was modified in order to achieve increased convergence speed based on a better exploitation of the search space. That is obtained especially through the way in which the offspring resulting from recombination enters the population. It does not take the place of its first parent, but that of the least fit individual in its replacement radius. Therefore, weak individuals are removed more aggressively (doubling the effect of merging) from the current population. For this reason, the stepping stone principle does not hold here anymore, but n (where n is the population size) random individuals are selected instead. Now individuals may be replaced by some offspring without ever being selected for reproduction. An important aspect of the method is the choice of the replacement radius value. If picked properly, this new parameter may lead to improved convergence speed.

The offspring obtained after recombination may replace an individual that does not belong to any of the mating regions of its parents, thus performing faster separation of individuals into clusters.

As the effect of quasi-generational survival selection, used by the classical GC, can be noticed only when another stepping stone does not find the initial individual in its mating region but meets the offspring instead, the new technique is totally generational. This means that the offspring that replaces its parent might be selected for reproduction many times in the same generation or might vanish within that iteration of the algorithm. Thus, the generational scheme leads to increased exploitation.

4.4 Genetic Chromodynamics for Classification

We recall the definition of a possible solution to the classification problem, imagined as a set of rules (class prototypes) that contains at least one rule for each class (Chap. 3, Sect. 3.11).

An individual is thus considered to be an IF-THEN first order logic entity in conjunctive form, holding the thresholds for the attributes in the data set, all referred by equality signs. An evolved prototype is thus a representative for a class of the problem and contains the attribute thresholds that place a sample into that class and prevent it from being labeled otherwise.

EAs naturally lead to a homogenous set of solutions, so the establishment and the preservation of diversity within the collection of the evolved rules is an intricate task that adds complexity to existing evolutionary classifiers. However, multimodal evolutionary techniques, such as GC, possess various intrinsic means for a longer maintenance of diversity. It guarantees that, in the end of the evolution process, each class will be covered by at least one prototype.

It is then reasonable to investigate whether crowding GC is also able to perform a simultaneous evolution of prototypes for the different outcomes. In this respect, we subsequently present the tailoring of the approach for classification as in [Stoean et al, 2005]. In short, a prototype population, covering all class labels, is evolved. Each individual is adjusted to be representative of the training samples with the same outcome. Once the set of final individuals is reached, each test sample is labeled by the proximal decision prototype.

We expect that several benefits arise from the employment of GC for classification:

- The encoding is simple to evolve and apply;
- Distinct prototypes for each class are implicitly maintained, in correspondence to the various colors of GC;
- It puts forward an economical EA-based solution to the task of following many optima in parallel.

4.4.1 Representation

Each individual c expresses a prototype for a class and consists of threshold values for each indicator of the classification problem and ends with the targeted label.

Hence, the formal expression of a prototype can be written as (4.3), where there exists at least one such rule for every class y_i, i=1,2, ..., k, of the problem. A threshold is a real valued number that denotes the center of a region marking the boundary for an attribute. Its combination to those of other features discriminates the samples into classes.

$$c : \text{IF } x_1 = threshold_1 \ldots \text{ AND } x_n = threshold_n \text{ THEN } class = y_i \qquad (4.3)$$

The values for the genes of all individuals are randomly initialized following a uniform distribution in the definition intervals $[a_i, b_i]$ of each corresponding attribute i of the data set. The class value for each individual is also randomly established from the set of possible outcomes. This representation is similar to that of the Michigan approach (Sect. 3.11).

4.4.2 Fitness Evaluation

The prototypes are evolved against the training set. The fitness of an individual is computed as the sum of distances to all training samples that have the same outcome as the individual.

To detail this, each individual has an outcome, which is the class label. To determine the fitness, one looks at training samples that have this class label. Then, the fitness is the sum of all distances, where each distance is calculated between the vector of thresholds and the vector of attribute values of each corresponding sample.

The aim is to minimize distances, conceiving thus good discrimination rules. The distance can be chosen as the common Euclidean measure for continuous attributes, for instance.

4.4.3 Mating and Variation

The mate for every individual is selected within its predefined mating region by proportional selection and must have the same class label. Intermediate recombination and mutation with normal perturbation are selected as variation operators. Mutation does not apply to the gene representing the outcome. For more information concerning the behavior of these operators see Chap. 3 (Sect. 3.7, 3.8 and 3.9). Note however that the choices for the operators are not the only ones that could be viable, but only those that have been tested as being more common for the given representation.

4.4.4 Merging

The merging operator does not take into account the outcome. Now, it is possible, in the early generations, that very similar individuals have different outcomes. As the fitness evaluation takes into account the quality of the prototype for the classification task, only the individuals with the proper label for the given weights will survive.

4.4.5 Resulting Chromodynamic Prototypes

After a fixed number of evolutionary iterations has elapsed, the final population contains at least k prototypes, one for each class of the problem. If there is more than one prototype for a certain category then, when applying them in order to label a new example, at least one needs to be satisfied for the sample to be labeled with that class.

Once a test example is available, its distance to all prototypes in the final population is computed. The individual which is closest to the test instance gives its predicted outcome. The accuracy is then computed as the number of successful label matches to the real targets of test samples over their total number.

4.5 Experimental Results

The crowding GC classifier [Stoean et al, 2005] is tested and applied for two data sets from University of California at Irvine (UCI) repository [Bache and Lichman, 2013]: Pima diabetes and iris. We want to investigate whether GC manages to maintain more distinct niches as the number of classes increases, i.e., from 2 classes in Pima to 3 classes in iris. The accuracies achieved for Pima diabetes and iris after 30 runs of crowding GC are of 69.52% and 92.84% with the standard deviations of 1.72% and 1.53%, respectively. The number of resulting prototypes for Pima diabetes is 2 and for iris is 3, therefore equal to the number of classes of each problem.

The ratio between the training and test sets is set to 2/3 training - 1/3 test samples and random cross-validation is performed 30 times. Parameters specific to GC (mating, merging and replacement radii) and those typical of the EA are set through the semi-automated tuning method of Sequential Parameter Optimization (SPO) [Bartz-Beielstein, 2006]. The SPO builds on a quadratic regression model, supported by Latin Hypercube Sampling (LHS) [Montgomery, 2006] methodology and noise reduction, by an incrementally increased repetition of runs.

In many tests for diabetes, a higher accuracy of 80% is obtained when the individual pool still has four prototypes left and has not converged yet. This leads to the idea that in the structure of each of the two obvious clusters two other subclusters are included. Thus, with a reported best (over all generations) instead of last accuracy, better results can be obtained. The same situation occurs for iris, where the best prediction accuracy across the evolutionary process was of 98%.

4.6 Concluding Remarks

GC creates and preserves subpopulations, each corresponding to a global/local optimum of the task, through the use of a stepping stone search mechanism and a local interaction principle, as selection for reproduction. Selection for replacement takes place between the resulting offspring and the current stepping stone individual. A merging operator is used to achieve decrease in the number of individuals.

In preliminary theoretical research, a crowding procedure was suggested for the integration of the offspring obtained after recombination but this also changes the way individuals are selected. Now they cannot each be considered in turn anymore, like in the original GC, because some might vanish prior to that, but they are instead randomly selected from the continually changing population. The goal of the technique was to speed up evolution while maintaining the ability of GC to find and preserve the global/local optima.

The crowding GC approach was tailored for addressing the classification task discussed throughout this book. Individuals represent attribute thresholds, while minimizing distances between them and the identically labeled training samples leads to good prototypes for each class, representing cluster centers in the n-attribute variable space.

As our first attempt of a plain multimodal EA to be transformed for classification, the results were not disappointing. The obtained advantages were:

- The maintenance of prototype diversity as connected to each class of the problem.
- The merging of similar individuals towards the achievement of a concise final rule set.
- Runtime equal or even better as opposed to that of an unimodal EA, due to the merging operator.

The drawbacks can be nevertheless noticed as the following:

- Low accuracy as compared to approaches designed for classification.
- The rule set is dramatically reduced in some iterations before convergence, leaving less individuals to express the different prototypes within the same class.
- The lack of collaboration between rules of different outcomes during the evaluation process towards a superior overall prediction accuracy.

Chapter 5
Cooperative Coevolution

I would like to see anyone, prophet, king or God, convince a thousand cats to do the same thing at the same time.
Neil Gaiman

5.1 Goals of This Chapter

The natural evolution of individuals implies adaptation to the environment, part of which consists of other living beings, in particular different groups or species. From this viewpoint, evolution is actually coevolution. Coevolution can be competitive, cooperative or both. Similarly, in the evolutionary computational area, EAs have been also extended to coevolutionary architectures. These are interesting because they define the fitness evaluation of an individual in relation and adaptation with respect to the other individuals in the population.

As one of the main focuses of this book is a cooperative approach for classification [Stoean et al, 2006], [Stoean and Stoean, 2009a], [Stoean et al, 2011a] this chapter will discuss only the collaborative side of coevolution. Its main components are therefore subsequently described, together with the standard application of the technique for function optimization [Potter and de Jong, 1994], [Wiegand, 2003]. In the second part, we will concentrate on its design for classification, where it is meant to evolve class prototypes that cooperate towards an accurate discrimination between samples.

The structure of the chapter is designed to contain the basic description of cooperative coevolution (CC) in Sect. 5.2 and its altering to address classification (Sect. 5.4), with a recollection of other such approaches that imply coadaptation (Sect. 5.3). Several enhancements triggered by experimentation (Sect. 5.5) are outlined in Sect. 5.6 and 5.7.

5.2 Cooperation within Evolution

Artificial coevolution is inspired from the interactive process that occurs between species in nature. On the one hand, species have to fight for the same resources, when they are in competition for a certain goal, and, on the other hand, different species collaborate for a specific purpose. As such, there are also two kinds of computational coevolution, distinguished from each other as follows. In competitive

C. Stoean and R. Stoean, *Support Vector Machines and Evolutionary Algorithms for Classification*, Intelligent Systems Reference Library 69,
DOI: 10.1007/978-3-319-06941-8_5, © Springer International Publishing Switzerland 2014

coevolution, the evaluation of an individual is determined by a set of competitions between itself and other individuals, while, in CC, collaborations between a set of individuals are necessary in order to evaluate one complete solution.

When solving a task by CC, a reasonable decomposition of the problem statement into components can be firstly achieved. Then, each subproblem is assigned to a population (or species). Every population evolves concurrently with the others, exchanging information only at fitness evaluations. Each individual in a population represents a piece of the solution to the problem and thus a potential candidate for every component in turn cannot be assessed apart from those of the complementary parts. Therefore, every individual of each species is evaluated by selecting collaborators from every other population. A complete solution to the problem at hand is thus formed and its performance is computed and attributed as the fitness value of the individual in turn to be evaluated.

CC is introduced as an alternative evolutionary approach to function optimization [Potter and de Jong, 1994]. For this task, one considers as many populations as the number of variables of the function, where each variable represents a component of the solution vector and is separately treated using any type of EA. Several functions with multiple local optima and one global optimum had been considered and the cooperative coevolutionary algorithm proved to be effective [Potter and de Jong, 1994], [Wiegand, 2003].

A simple example of a function optimization illustrates the early use of the algorithm in Fig. 5.1. Consider the Schwefel function for 3 variables (described in (5.1)). Three corresponding populations, one for each variable, are filled with individuals and these are evolved separately. Collaboration takes place when fitness is evaluated as shown in Fig. 5.1.

$$f(x,y,z) = x \cdot sin(\sqrt{|x|}) + y \cdot sin(\sqrt{|y|}) + z \cdot sin(\sqrt{|z|}), \quad \text{where } x,y,z \in [-500,500] \tag{5.1}$$

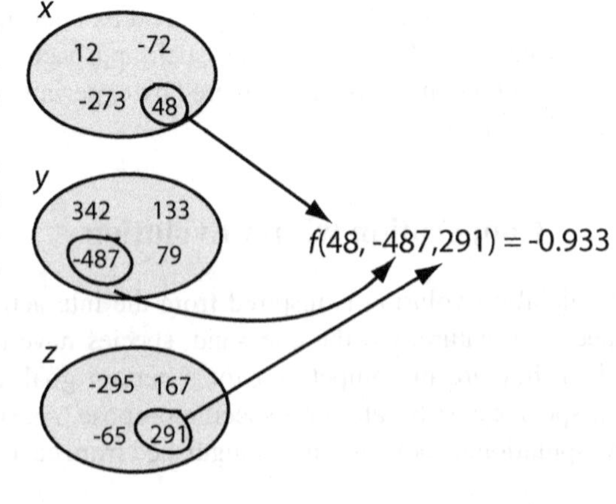

Fig. 5.1 Evaluation of a complete solution for Schwefel function with 3 variables (described in formula (5.1)). When evaluating an individual from population x, e.g. 48, collaborators from populations y and z are selected to form a complete solution that can be evaluated in the function expression.

Algorithm 5.1 outlines the general cooperative coevolutionary algorithm. It starts with the initialization of each population. In order to measure the quality of a certain individual for the first evaluation, a random selection of individuals (collaborators) from each of the other populations is performed and obtained solutions are evaluated. After this starting phase, each population is evolved using a canonical EA. For the next evaluations, there are three attributes that govern collaborator selection [Wiegand et al, 2001]:

1. *Collaborator selection pressure* refers to the way of choosing for the individual under evaluation each collaborator from every other population in order to form a complete solution to the problem. One can thus select the best individual according to its previous fitness score, pick a random individual or use classic selection schemes.
2. *Collaboration pool size* (cps) represents the number of collaborators that are selected from each population.
3. *Collaboration credit assignment* decides the way of computing the fitness of the current individual. This attribute appears when the collaboration pool size is higher than 1. There are three possibilities for calculating the fitness value based on multiple collaborators of an individual:

    a. *Optimistic* - the fitness of the current individual is the value of its best collaboration.
    b. *Hedge* - the average value of its collaborations is returned as its fitness score.
    c. *Pessimistic* - the value of its worst collaboration is assigned to the current individual

Although in the experiments driven for function optimization [Wiegand et al, 2001], both the pessimistic and hedge strategies consistently resulted in significantly poorer performance, they proved to be in fact effective in our approach for classification in Sect. 5.4.

Presuming once more that a maximization problem is at hand, Algorithm 5.2 outlines the fitness evaluation manner of CC with respect to the three mentioned attributes. In order to evaluate an individual c from a certain population, a number of complete potential solutions are formed according to the chosen collaboration pool size. In order to aggregate a solution, collaborators from each population different from that of c are selected through a certain strategy (collaboration selection pressure). Each solution is evaluated according to the objective function of the current problem. Once all candidate solutions are gathered and assessed, the preferred type for the collaboration credit assignment decides the value that will be returned as the performance of individual c. Naturally, the three attributes regarding the selection of the collaborators are set offline (prior to the evolutionary process) and are not changed later in its course.

Since a fitness evaluation is measured as a team result, the performance of a candidate solution largely depends on the chosen collaborations. Different collaborators may conduct to distinct outputs, so various compatibilities between individuals from

Algorithm 5.1 The CC algorithm.

Require: An optimization problem broken down into components
Ensure: The set of subsolutions that make up together a complete solution to the problem
  begin
  $t \leftarrow 0$;
  for each species s **do**
    Randomly initialize population $P_s(t)$;
  end for
  for each species s **do**
    Evaluate $P_s(t)$ by choosing collaborators from the other species for each individual;
  end for
  while termination condition is not satisfied **do**
    for each species s **do**
      Select parents from $P_s(t)$;
      Apply genetic operators;
      Evaluate offspring by choosing collaborators from the other species for each individual;
      Select survivors in $P_s(t+1)$;
    end for
    $t \leftarrow t+1$;
  end while
  return a complete solution
  end

Algorithm 5.2 Fitness evaluation for an individual c within CC.

Require: A current individual c
Ensure: The fitness evaluation of c
  begin
  for each $i = 1, 2, ..., cps$ **do**
    select one collaborator d_j, $j = 1, 2, ...,$ *number of species* from each population different from that of c;
    form a complete potential solution;
    compute the fitness f_i of the solution in the terms of the objective criterion;
  end for
  if *Collaboration credit assignment* = *Optimistic* **then**
    $evaluation \leftarrow max_{i=1}^{cps}(f_i)$;
  else
    if *Collaboration credit assignment* = *Pessimistic* **then**
      $evaluation \leftarrow min_{i=1}^{cps}(f_i)$;
    else
      $evaluation \leftarrow avg_{i=1}^{cps}(f_i)$;
    end if
  end if
  return *evaluation*
  end

distinct species can appear, and thus diversity is maintained for a longer period of time.

Having the general CC framework outlined, we will now proceed to its tailoring for our classification set task.

5.3 Evolutionary Approaches for Coadaptive Classification

There are several examples of EA techniques that imply coadapted components:

The Michigan approach [Holland, 1986]: cooperation is achieved by the bucket brigade algorithm that awards rules for collaboration and penalizes them otherwise (Chap. 3, Sect. 3.11).

The REGAL system [Giordana et al, 1994]: EAs evolve stimulus-response rules in conjunctive form, similar to those representing prototypes in the approaches gathered in this book. Problem decomposition is performed by a selection operator, complete solutions are found by choosing best rules for each component, a seeding operator maintains diversity and the fitness of individuals within one component depends on their consistency with the negative samples and their simplicity.

GC [Stoean and Dumitrescu, 2006]: an EA technique that achieves multiple optima search through the use of radii can be also employed for classification (Chap. 4).

As concerns the more recent discussion upon the cooperative coevolutionary technique, it itself had also been successfully applied to develop a rule-based control system for agents. Two species were considered, each consisting of a population of rule sets for a class of behaviors [Potter et al, 2001].

To the best of our knowledge, there has been no other previous attempt in applying CC to classification based on individuals that encode simple conjunctive IF-THEN prototypes in first order logic, holding threshold information on attributes.

5.4 Cooperative Coevolution for Classification

Recall the GC approach for classification in Chap. 4, where multiple class threshold prototypes were maintained through mating restrictions and converged to a decision set by a merging scheme. We currently want to put forward a simple evolutionary classification technique similar to that, but additionally accomplish the collaboration between evolving prototype populations of different outcomes towards an enhancement in prediction accuracy. A potential architecture considers the final decision output of CC as to contain for each outcome of the classification task one or more prototypes selected randomly from the ending populations.

As a result, the decomposition of the problem into subtasks is performed by assigning to every species (population) the job of building the prototype(s) for each class [Stoean et al, 2006], [Stoean and Stoean, 2009a], [Stoean et al, 2011a]. Thus, the number of species equals the number of outcomes of the classification problem.

Usually, classification tasks do not possess a high number of classes, so the number of species will not be too large, resulting into a reasonable runtime for CC.

An example of application of the algorithm to Fisher's iris data set is illustrated in Fig. 5.2. The data set contains 4 attributes described on the top of the figure and 3 classes: *setosa*, *virginica* and *versicolor*. Therefore, 3 populations are considered, one for each class. In order to evaluate a certain prototype from a population, a complete set of rules has to be formed by gathering individuals from the all other populations as collaborators. The set is afterwards applied to the training data and the accuracy is assigned as the fitness of the initial rule. More details are given in the following subsections.

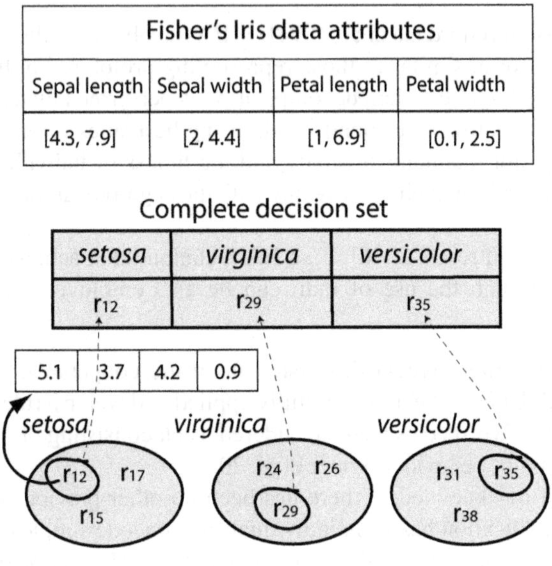
Fisher's Iris data attributes
Sepal length
Sepal width
Petal length
Petal width
[4.3, 7.9]
[2, 4.4]
[1, 6.9]
[0.1, 2.5]
Complete decision set
setosa
virginica
versicolor
r12
r29
r35
5.1
3.7
4.2
0.9
setosa
virginica
versicolor
r12
r17
r15
r24
r26
r29
r31
r35
r38

Fig. 5.2 Three populations evolve prototypes, each for one of the three classes of Fisher's iris data set (4 attributes, 3 classes). When evaluating a rule (such as r_{12}), collaborators are selected from all the other populations for achieving a complete decision set that can be further applied to the data. Four particular threshold values of the r_{12} prototype, corresponding to the four attributes, are also illustrated for example.

Regarding previous attempts in literature and the CC particularities among EAs, here are thus the advantages of a CC approach to classification (even as compared to GC):

- The representation is once more in a simple IF-THEN equality format with thresholds for attributes.
- Cooperation is implicit from the CC construction.

- Subpopulations connected to every class are inherently maintained by the CC.
- Different means to collaborate can be easily exploited.

As the task of the CC technique is to build k prototypes, one for each class, k populations are therefore considered, each with the purpose of evolving one of the k individuals.

5.4.1 Representation

As earlier within the GC model for classification (Chap. 4), each individual c is formed by conjunctions of attribute thresholds conducting to a class of the problem (recall formula (4.3)). Within the cooperative approach to classification however, an individual will not encode the class anymore, as all individuals within a population have the same outcome.

5.4.2 Fitness Evaluation

When measuring the representative quality of a potential prototype, it has to be integrated into a complete set of solutions which is subsequently applied to the training set (Fig. 5.2). The obtained accuracy reflects the quality of the initial individual. As earlier mentioned, the value of the accuracy very much depends on the other rules that are selected. For a more objective assessment of its collaborative quality, the prototype is tested within several distinct sets of solutions, i.e. different values for the collaboration pool size (cps) are considered.

For evaluating an individual c from a certain population – that is a prototype for a certain outcome – a collaborator from each of the other populations is selected. The selection is carried out n times, according to the collaborator selection pressure choice. Every time, the set of rules is applied to the entire training collection. When it is decided which prototype is closer to a training data point, the distance between the selected individuals and the sample has to be computed. The obtained accuracy represents the evaluation of the initial individual c and it is computed as the percentage of correctly classified samples from the training set. The fitness of c may be then given by one of the cooperative assignments (optimistic, hedge or pessimistic) for the cps acquired accuracies. A description of these possibilities of evaluating the fitness for an individual c is given in Algorithm 5.3.

A nonstandard type of assignment was additionally introduced in [Stoean and Stoean, 2009a] and is outlined by Algorithm 5.4. For each sample s in the training set, multiple sets of rules are formed and applied in order to predict its class. All prototypes within a set have different outcomes. For sample s, scores are computed for each of the possible outcomes. When a set of prototypes is applied to a sample, a certain outcome is established for the latter, given by the class of the prototype of highest similarity. The score of that outcome is increased by unity. Each of the cps sets of rules are applied to s. Finally, the class of s is concluded to be the class that obtains the highest score.

Algorithm 5.3 An individual c encoding a class prototype is evaluated by means of either optimistic, pessimistic or hedge collaboration credit assignment within CC for classification.

Require: A current individual c
Ensure: The fitness evaluation of c by CC
begin
for $i = 1$ to cps **do**
  $correct_i = 0$;
  Select a random collaborator from each population different from that of c according to the collaborator selection pressure parameter;
  for each sample s in the training set **do**
    Find prototype r from the set of all collaborators that is closest to s;
    Class for s becomes the class of r;
    if the found class for s equals the real class of s **then**
      $correct_i \leftarrow correct_i + 1$;
    end if
  end for
end for
if *optimistic* **then**
  $success \leftarrow max_{i=1}^{n}(correct_i)$
else
  if *pessimistic* **then**
    $success \leftarrow min_{i=1}^{n}(correct_i)$
  else
    $success \leftarrow avg_{i=1}^{n}(correct_i)$
  end if
end if
$accuracy \leftarrow 100 * success$ / number of training samples;
return *accuracy*
end

It may happen that, for a certain sample, there exist more classes that have the same highest score. In this case, one class still has to be decided and we choose the first one in the order of outcomes. As all combinations of rules count in the determination of accuracies, we can state that the new choice of assignment is closer to the classical hedge type.

5.4.3 Selection and Variation

As before, resorting to the common choices for EA operators, we employ tournament selection, intermediate recombination and mutation with normal perturbation (see Chap. 3 for more details regarding these types of operators).

Algorithm 5.4 A score-based fitness evaluation for an individual c within CC for classification.

```
Require: A current individual c
Ensure: The score-based fitness evaluation of c by CC
  begin
  for each sample s in the training set do
     Set the score for each possible outcome of s to 0;
  end for
  for i = 1 to cps do
     Select a random collaborator from each population different from that of c according to
     the collaborator selection pressure parameter;
     for each sample s in the training set do
        Find the rule r from the set of all collaborators that is closest to s;
        For s, increase the score for class of r by one unit;
     end for
  end for
  success ← 0;
  for each sample s in the training set do
     if the real class of s equals the class of highest score for s then
        s is correctly classified;
        success ← success + 1;
     end if
  end for
  accuracy ← 100 * success / number of training samples;
  return accuracy
  end
```

5.4.4 Resulting Cooperative Prototypes

After a preset number of EA iterations is reached, there are p populations of discovered prototypes. In order to form a complete final set, an individual from each resulting population has to be chosen. The selection may be carried out randomly, in an elitist fashion or by means of a standard EA scheme for this operator, taking into account the final fitness evaluations. However, selecting the fittest rule from each population does not necessarily trigger the best accuracy on the test set. Even if these best rules gave very good results on the training set, they may be in fact not general enough to be applied to previously unseen data (overfitting).

In order to acquire a measure of the algorithm performance for classification, the accuracy is computed in the following manner. For a number of cps times, one rule from each population is randomly selected in order to form cps complete sets of prototypes. Each time, the resulting set is applied to the test data in a similar manner to the fitness calculation that is computed with respect to the training set in Algorithm 5.4. Following thus the same voting mechanism, test success rate is eventually obtained.

5.5 Experimental Results

Two benchmark classification problems from UCI are considered for the early validation and comparison experiments: Wisconsin breast cancer diagnosis and iris recognition. The former is a two-class instance, while the latter represents a multiclass task, which should reveal whether the CC remains flexible and feasible with some increase in the number of outcomes. Recall that an increase in the number of classes determines a corresponding raise of the number of species.

Additionally, a real-world data set, dealing with hepatic cancer early and non-invasive detection, is also considered. It was brought to our attention in the past, with the purpose of delivering an efficient and comprehensible model of computer aided diagnosis. It moreover serves as a good test and application of CC on an unpredictable environment that is usually associated with raw data. Data is collected from the University Hospital in Craiova, Romania and contains 299 records of patients described by 14 significant serum enzymes and two possible outcomes (presence and absence of the disease).

Preliminary testing had been carried out in order to investigate different settings for the coevolutionary parameters for the breast cancer and iris data sets. The several choices for the collaboration credit assignment trigger no significant difference in resulting accuracy. Nevertheless, slightly better outcomes appeared to be achieved for the score-based fitness. When the collaboration pool size value is increased, the runtime of the algorithm expectedly also raises. The results also seem to be improved to some extent by the increase of the value for this parameter. The technique had been tested from one up to seven collaborators. For the collaborator selection pressure parameter, random selection drives the coevolutionary process to accurate results. If, instead, the best individual from each population is chosen for collaborations, the obtained results are surprisingly worse than when it is randomly picked. Finally, if proportional selection is employed for choosing the collaborators, results are slightly better than those obtained through random selection.

In order to choose the appropriate EA parameters (of the variation operators), the SPO approach [Bartz-Beielstein, 2006] is used once more. 30 runs were conducted for each data set; in every run, 2/3 random cases were appointed to the training set and the remaining 1/3 went into test.

The results obtained by CC on the three data sets are outlined in Table 5.1. A thorough statistical comparison of its outcomes against those of the top performing SVMs will be presented in the chapters describing their hybridization.

Table 5.1 The mean accuracy and standard deviation after 30 runs of CC on the chosen problems

Data set	Average accuracy (%)	Standard deviation (%)
Breast cancer	95.4	1.56
Iris	96.3	1.8
Hepatic cancer	90.1	2.67

We check the speed of convergence of the CC algorithm by applying the evolved prototypes to the test set from an early stage of the evolution process. This test is performed solely on the breast cancer data set. In the initial (approximately 10) generations, the results are quite hectic, jumping from 20% to 80% and in-between, reaching then a certain stability of about 80% accuracy and eventually growing slowly but constantly. In the final generations, there are only minor modifications of the test accuracy of up to one percent. However, the results greatly depend on the way the training/test sets are generated as there are cases when accuracy starts at 80% even from the early iterations.

As concerns the ease of parametrization of the CC, there is not a very strong dependence between the common EA parameters and obtained results, as very competitive accuracies are obtained for a large scale of their values. This is yet another argument in favor of the CC, as parameter setting is a delicate issue within EAs.

5.6 Diversity Preservation through Archiving

If multiple distinct prototypes for a class could be maintained within each population, the CC would cover more of the sample space particularities for each outcome. Since the regular EA that controls every population conducts to the generation of prototypes around one singular global/local optimum, the technique should be endowed with a mechanism to conserve a diverse selection of individuals. Further research therefore implied the creation of a concurrent archive that is constantly updated at the end of every evolutionary iteration by seizing a preset number of best collaborations from the current population and its previous content. These fittest combinations are copied into the new archive such that there are no two prototypes alike.

The enhanced version [Stoean and Stoean, 2009a] is outlined by Algorithm 5.5 and fundamentally follows the initial approach with the additional consideration of an external population of size $k * N$ as archive. k is the number of classes of the problem and N is a parameter of the method that stands for the number of rule sets to be at all times stored. The archive is initialized with the fittest prototypes, as derived from the first evaluation of the individuals in each population. The collaborators for each of these rules are also captured into the archive. At the end of every generation of the evolutionary loop, the best sets (with one prototype per class) from the present population and previous cached collections are conserved into the current archive. In the end of the algorithm, the archive consequently contains a multiple, diverse decision set. All the prototypes contained in this final archive are then applied to the test samples and the prediction accuracy is computed.

During pre-experimentation, it was noticed that, from the early stages of evolution, the archive had been populated with copies of the same best combination of individuals, so it was necessary to check whether those that are stored are identical or not. By allowing only different sets of individuals to become members of the archive, we assure a good diversity among the final set of prototypes of the external population. The proposed add-on demonstrated an enhancement of the

Algorithm 5.5 The CC classifier with an archive.

Require: A classification problem with k outcomes
Ensure: The archive holding multiple prototypes for each class
begin
$t \leftarrow 0$;
$P(t) \leftarrow \Phi$
for each class j **do**
Randomly initialize $P_j(t)$;
$P(t) \leftarrow P(t) \cup P_j(t)$;
end for
for each class j **do**
Evaluate $P_j(t)$ with collaborators from each other subpopulation for every individual;
end for
Copy the best N collaborations from $P(t)$, based on previous evaluation scores and neglecting identical rules, to archive $A(t)$;
while termination condition is not satisfied **do**
for each class j **do**
Select parents from $P_j(t)$;
Apply genetic operators;
Evaluate offspring with collaborators from each other subpopulation for every individual;
Select survivors in $P_j(t+1)$;
$P(t+1) \leftarrow P(t+1) \cup P_j(t+1)$;
end for
Copy the best N collaborations from $P(t+1)$ and $A(t)$ to $A(t+1)$;
$t \leftarrow t+1$;
end while
return the archive $A(t)$
end

methodology even from the early generations, when the confrontation of the individuals in the archive with the training set yielded good accuracies. This happens due to the fact that the archive contains the best found (up to the moment) suites of rules that match the known data and are, at the same time, different from each other.

Within the testing stage, we maintained the same coevolutionary parameters, as well as the same selection, variation operators and evolutionary parameters for the CC strategy with an archive as before. The value for the new parameter that refers to the size of the archive had been varied from 2 up to 10. Given the manner in which we test the archive against the given data, i.e., apply all stored rules to the test samples, this parameter cannot be very high both due to a subsequent increase in runtime and because a large archive does not necessarily exhibit significantly better results, as also noticed in the undertaken experiments. The accuracies reported by this variant were obtained for an archive size equal to 9 in case of breast cancer, 2 for iris, while for hepatic cancer the value was set to 5. The outcome of experimentation argues in favor of the archiving strategy. The results for each problem were compared via a Wilcoxon rank-sum test. The p-values (0.05 for breast cancer, 0.06 for iris and 0.31 for hepatic cancer) suggest to detect a significant difference only for the UCI data sets.

The accompanying repository thus manages to lead to some relevant improvement for the CC classification technique. The explanation could be that the archive already contains a collection of rules that are well-suited to the given data and as a consequence, no collaborations need to be further selected in the test stage, as in the primary construction. Moreover, during evolution, one rule can find collaborators and, in conjunction, obtains a promising training accuracy, but afterwards, when other collaborators are selected for the same individual, they jointly might output poor results. The EA would, in such a case, encourage the modification of these individuals (through mutation and/or recombination) or, even more likely, their removal through the selection operator and, as a consequence, they would be lost. The archive however collects these promising combinations and saves them from becoming extinct. The reported results underline the fact that the archiving strategy generally performs better than the standard CC approach.

5.7 Feature Selection by Hill Climbing

Providing only threshold prototypes alongside prediction may not be enough for real-world tasks like automated disease diagnosis. For instance, a problem of correct evaluation of the degree for hepatic fibrosis was in need of more explanations following the computer decision, such as highlighting the most important features that can be combined in order to give more accurate results. The chronic hepatitis C data set came from the 3rd Medical Clinic, University of Medicine and Pharmacy, Cluj-Napoca, Romania. It consists of 722 samples, each described by 24 indicators and 5 possible outcomes [Stoean et al, 2011a].

Techniques for automated diagnosis in medicine usually employ some mechanism of selecting the most relevant indicators in the data set prior to the classification [Shen et al, 2009], [Marinakis et al, 2009]. It is assumed that some attributes might only hinder the search for the accurate solutions or even submit the entire method to the curse of dimensionality. Moreover, when human physicians reach a diagnosis in practice, an effective rule to achieve a prediction cannot logically make reference to absolutely all attributes of the data. In this way, in the end the expert can see solely the essential indicators together with their thresholds and thus get a fast and reliable assistance in diagnosis. Alternatively, having all indicators present, even though the values of some may be disregarded for the specific outcome, cannot but harden and make the reading of the decision statement confusing.

Initially, CC was tried on the entire fibrosis data set directly, without any prior feature selection. The average results over 30 repeated runs only reached 51.8% test accuracy. This fact convinced us furthermore to support CC through an additional mechanism that would preprocess the data set. One of the commonly used feature extraction mechanisms, principal component analysis, was first employed in the study. This led to a sizeable reduction of data dimensionality from 24 to only 6 attributes. When applying CC to the resulting data, the results were only improved by 1.4%, as opposed to those achieved through the direct application on the original data. It has to be mentioned that there was however a major improvement in runtime.

The next choice was to use a GA for selecting the attributes. However, two GAs (including the CC) would have triggered an increase in runtime. In order to make the additional procedure efficient, a limited number of applications of the CC technique is nevertheless desired. For that reason, a hill climbing (HC) algorithm can be used with the purpose of choosing the attributes to make CC perform more efficiently.

The setup of the HC is performed as follows. An individual is represented through a binary encoding and its number of genes equals the number of features of the data set. When a gene has a value of 1, it means that the corresponding attribute is taken into consideration, while, when a value of 0 appears, it signifies that the attribute is skipped. Recall that, unlike EAs, a HC uses a single point movement through the search space. An individual is thus randomly constituted and the selected attributes are further fed to the CC. The approach generates prototypes based on the newly defined training set (where, for each sample, the attributes that have 0 values in the HC individuals are skipped) and then applies them to the test set (reformulated in the same way). A prediction accuracy is obtained and that value represents the evaluation of the HC. Perturbation is then applied for the individual, a new climber is obtained, it is evaluated and, if fitter, it replaces the previous one.

Now, turning to the experimental side, when evaluating a HC individual, the CC is applied 30 times to the data set referring only the chosen features. In each of the 30 runs, the training and test sets are randomly chosen in order to have a more objective evaluation. Strong mutation (see Chap. 3) is then applied for variation. The loop goes on until there is no improvement in fitness for a number of iterations (20), then a new individual is generated and the process restarts. A fixed budget of fitness evaluations (1000) is set for the HC as a stop condition for the algorithm.

The average accuracy, following the 1000 HC evaluations, reaches 55.93%. The individual that yields the best obtained test accuracy (62.11%) only selected 9 attributes out of the 24 available. It happens that the test accuracy even reaches above 65% correctly classified samples. One such example run of the CC is illustrated in Fig. 5.3.

Practically, there are several configurations of medical indicators found by the HC algorithm that produce good results. It is very interesting to observe that there are not always the same attributes that are selected. We can thus state that the HC algorithm rather discovers sets of features that in connection perform better. However, there are some attributes that are included more often into many successful configurations and, in order to discover them, best 3% of the 1000 different evolved configurations are examined and the importance of the features is illustrated in Fig. 5.4. It has to be nonetheless underlined that the most significant feature, the one that has been chosen in most of the successful combinations, was the liver stiffness indicator (attribute a1 in the figure), fact that is also acknowledged by the medical experts [Gorunescu et al, 2012].

Also, at this point of testing our CC approach on a real problem, we also decided to measure the dissimilarity between the individuals from the same species

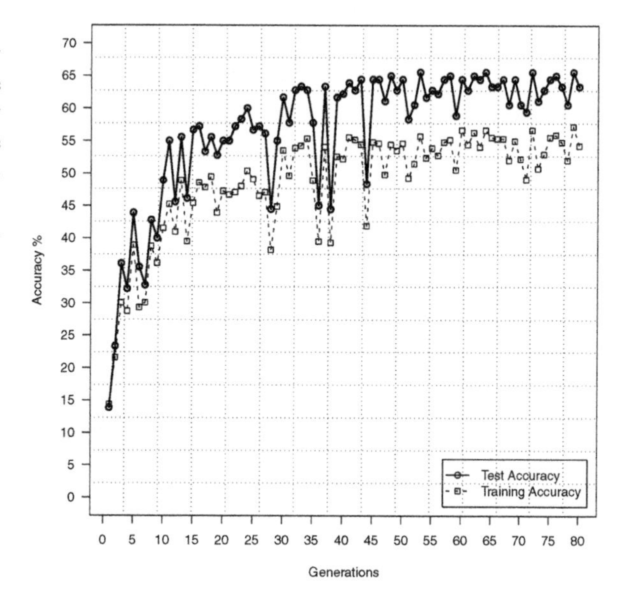

Fig. 5.3 Obtained accuracy results in a good run of the best configuration found by the HC algorithm. Both the training and test accuracies are illustrated from generation 1 and up to generation 80 [Stoean et al, 2011a].

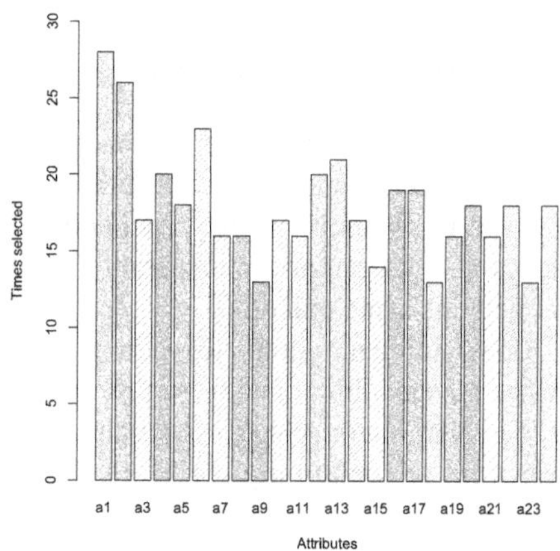

Fig. 5.4 The attributes more often selected (from 30 configurations) that yield the best results [Stoean et al, 2011a]

during the evolutionary process. If the technique is correct and each of the different populations converges to an optimum of the problem, then these distances should diminish as generations pass. Figure 5.5 shows that the distance between prototypes of each class decreases almost in the same manner. When individuals are randomly generated, all dissimilarities start from a large value and eventually converge to a comparable small value after 80 generations. The dissimilarity within a species is obtained by adding the distances from each individual to the mean of that population.

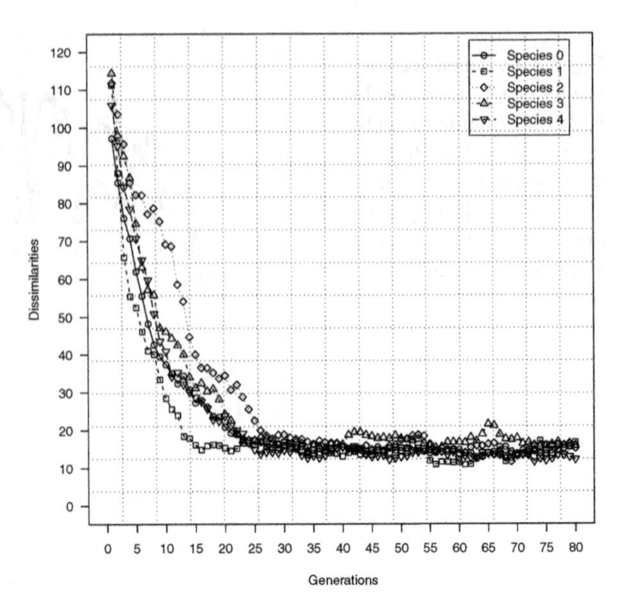

Fig. 5.5 Computed dissimilarities within each of the five populations for the same run as in Fig. 5.3 [Stoean et al, 2011a]

5.8 Concluding Remarks

This chapter described one approach to classification based on CC. Its workings are simple and yet efficient. Populations of prototypes with the same outcome are evolved in parallel and they cooperate during the evaluation process. The fitness assignment is designed to constrain the rules to resemble the samples in the training set with the same outcome.

Coevolutionary parameters additionally play a role towards better accuracy. The cooperative parameter with the highest influence on the final result of the algorithm is the collaborator selection pressure. A selection scheme is preferred to a random selection; selecting only the best collaborator seems to be the worst of the three choices.

The collaboration pool size parameter directly influences runtime. The final test accuracy is also affected by the increase in this value, but eventually a balance has to be established between runtime and accuracy. Another important observation is that, when a certain threshold for the collaboration pool size parameter is surpassed, there is no further gain in accuracy. We generally achieved the best results when three collaborators were considered.

Changes within the collaboration credit assignment do not bring vital modifications to the final average results. The differences between various settings as regards the final accuracies do not exceed one percent.

In order to boost accuracy, an archiving strategy is further on added to the CC. The mechanism assumes the conservation of the most promising sets of rules starting from the first generation. After each iteration of the EA, the archive is updated with the best performing associations of class prototypes from the current population and its own content. The prototypes of the final archive are in the end of the algorithm applied together to the test set. The results were according to the plan, as they proved to be more accurate than the ones acquired by appointing the primary cooperative classifier. The CC approach with an archive was built to record only suites that contain one rule per class. Nevertheless, storing sets that allow several rules per outcome might make the classifier even more versatile and efficient.

A HC subsequently searches for a possible boolean selection of attributes from the data set. Every candidate configuration is measured against the training set (holding only the selected indicators) by means of the CC. For the given selection of features, corresponding genotypic prototypes evolve within different species for each class. The performance of a resulting set is the evaluation credited to the HC formed combination of attributes. Besides a significant improvement in accuracy, the proposed mechanism additionally has the essential advantage that it reveals sets of attributes that, even if at first glance may not appear as the most suitable choice for the human expert, it is through association that they provide a better accuracy than when using the entire set of features. A dissimilarity computation has also been performed, in order to illustrate that each CC population eventually exhibits homogeneity in prototypes for a class.

Compared to the previous GC methodology, that is also designed for classification, the CC alternative benefits from the implicit maintenance of a number of subpopulations equal to the number of classes. It also inherently performs cooperation between rules of different outcomes, which is nonexistent in GC. Its drawback lies however in a higher runtime, due to more populations of individuals. Despite this, the accuracy is superior to that of GC, as it can be seen from the experiments. That is the reason why prototype evolution after SVM learning is performed in Chap. 7 only on the CC variant.

We can then outline the main advantages of CC in opposition to existing EAs for classification:

- A built-in mechanism for the maintenance of parallel evolution for class prototypes.
- More control on the coexistence of the different niches.
- An implicit engine for cooperation between all class prototypes, where different parameters drive the collaboration.

Acknowledgements. The authors wish to thank Prof. Adrian Saftoiu, MD, PhD, from the University of Medicine of Pharmacy, Craiova, Romania, for the data on hepatic cancer. We also thank Prof. Radu Badea, MD, PhD, Monica Lupsor MD, PhD, and Horia Stefanescu, MD, PhD, from the University of Medicine and Pharmacy, Cluj-Napoca, Romania, for the fibrosis data.

Part III
Support Vector Machines and Evolutionary Algorithms

The third and final part of this book presents possible combinations between SVMs and EAs that have been explored by the authors. After presenting the context of existing techniques for this purpose, two hybridizations are outlined:

- the ESVM, where their joining is performed at the level of an EA solving the SVM primal optimization problem,
- and the SVM-CC, where EAs extract the logic following SVM learning.

Chapter 6
Evolutionary Algorithms Optimizing Support Vector Learning

Nothing of me is original. I am the combined effort of everyone I've ever known.
Invisible Monsters by Chuck Palahniuk

6.1 Goals of This Chapter

This chapter outlines the structure and particularities of a recently developed EA engine to boost the flexibility of solving the PP within SVMs. The framework of the evolutionary-driven support vector machines (ESVMs) resolves the complexity of modeling the embedded optimization problem, opens the 'black box' of the solving component and broadens the spectrum of maneuvering the powerful kernel machine of SVM for classification tasks.

The methodology inherits the geometrical consideration of learning within SVMs. Alternatively, the estimation of the coefficients of the decision surface (2.2) is addressed through the direct search capabilities of EAs with respect to accuracy and generalization. It thus presents itself as a technique to target the primal optimization problem, in opposition to other recent EC approaches that either solve the DP [Mierswa, 2006a] or consider further SVM enhancements such as kernel evolution [Howley and Madden, 2004], parameter approximation [Friedrichs and Igel, 2004] or selection of best features [Kramer and Hein, 2009].

The novel algorithm ranks competitive when compared to the original SVM approach. Real-world problems from biology, medicine and Internet have been consecutively referred, showing the general, efficient nature of the proposed framework. It must be noted, however, that for those situations when SVMs with standard kernels can achieve the task, we recommend the use of the regular method from mathematical optimization to solve the problem - it is computationally cheaper and EAs are not needed to provide a similar result. However, if either non-standard kernels have to be utilized or insight into the optimization box is necessary, it is adequate to use the EA-based version.

While the ESVM utility for allowing reference to kernels of any form with no further constraints is obvious, the meaning and interpretation of their use in also outputting the decision coefficients may be unclear. The real advantage of the latter, apart from being able to follow the solving of the optimization problem, comes to one's attention again when attempting to solve a practical problem of automated

C. Stoean and R. Stoean, *Support Vector Machines and Evolutionary Algorithms for Classification*, Intelligent Systems Reference Library 69,
DOI: 10.1007/978-3-319-06941-8_6, © Springer International Publishing Switzerland 2014

medical diagnosis. The weights for such medical attributes show the importance of each within the decision-making process.

The chapter is shaped as follows. After looking at prior combinations between SVMs and EAs (Sect. 6.2), ESVMs are introduced in Sect. 6.3. Based on the experimental findings (Sect. 6.4), several enhancements tailored according to the found practical needs are presented in Sect. 6.5 and 6.6.

6.2 Evolutionary Interactions with Support Vector Machines

This is obviously not the first attempt to bring SVMs and EAs close together. Existing alternatives are numerous and recent, of which some are presented further on. Strict hybridization towards different purposes is envisaged through: model and feature selection, kernel evolution and evolutionary detection of the Lagrange multipliers.

- Model selection concerns the adjustment of hyperparameters (free parameters) within SVMs. These are the penalty for errors C and parameters of the kernel which, in standard variants, are determined through grid search or gradient descent methods. An EA generation of hyperparameters can be achieved through evolution strategies [Friedrichs and Igel, 2004].
- When dealing with high dimensional problems, feature selection is often employed for providing only the most relevant attributes as input for a SVM. The optimal subset of features can be evolved using GAs [de Souza et al, 2005] and genetic programming [Eads et al, 2002].
- Evolution of kernel functions to model training data is performed by means of genetic programming [Howley and Madden, 2004].
- The Lagrange multipliers involved in the expression of the DP can be evolved by means of evolution strategies and particle swarm optimization [Mierswa, 2006a].
- Inspired by the geometrical SVM learning, the paper [Jun and Oh, 2006] reports the evolution of w and C while using erroneous learning ratio and lift values inside the objective function.

The ESVM approach, however, focuses on the evolution of the coefficients of the decision function as they appear within the PP of the SVM. To the best of our knowledge, a similar research had not been accomplished yet.

6.3 Evolutionary-Driven Support Vector Machines

The ESVM framework [Stoean et al, 2009a], [Stoean et al, 2009b] views artificial learning from the SVM geometrical perception and determines the coefficients of the decision hyperplane through a standard EA. A hyperplane geometrically discriminates between training samples and its equation must be approximated, with respect to both the particular prediction ability and the generalization capacity. Individuals of an evolutionary population therefore encode the hyperplane coefficients and they evolve to optimally satisfy the SVM objective and constraints towards an accurate separation into classes.

In Chap. 2 we have seen that the optimization statement that is reached by SVM learning may be standardly resolved by relying on a mathematically complex extension of the Lagrange multipliers technique [Vapnik, 1995b]. A dual formulation is derived and the optimal Lagrange multipliers are considered as the solutions of the system resulting by setting the gradient of the new objective function to zero. Once the Lagrange multipliers are found, several conditions may be used to further compute the coefficients of the hyperplane.

Nonetheless, existing SVM implementations (such as LIBSVM [Chang and Lin, 2001] or SVM light [Joachims, 2002], for instance) implicitly output only the target for a test case (or the mean accuracy, if several samples have to be labeled) as derived from the black box training. The formula describing the relationship among indicators that led to that decision cannot be straightforwardly visualized. In this respect, the optimization task may be alternatively plainly addressed by the adaptable and general EAs that are able to inherently and, what is more, unrestrainedly determine the coefficients that lead to an optimal decision.

6.3.1 Scope and Relevance

There are several aims that led to the appearance and design of the ESVM framework and they come from the disadvantages that the standard SVM paradigm brings along to a modeler or user:

- Despite the originality and performance of the learning vision of SVMs, the inner training engine is intricate, constrained, rarely transparent and able to converge only for certain particular decision functions. This has brought the motivation to investigate and put forward an alternative training approach that benefits from the flexibility of EAs.
- Although learning is fast, present SVM implementations do not implicitly provide a formula for parameters-decision interaction, which is extremely useful for gaining inside information into the practical phenomenon. What is more, such a formula does not give only information on the interplay between attributes and responses, but also fast support for an immediate reaction from the system when confronted with a novel test case. And this proves vital when including such a decision support for medical diagnosis.
- Although these decision equations could be nevertheless computed within SVMs, this requires further inspection of the available implementations and the design of a second application that could store them and be called whenever a new case appears. The SVM algorithm may be implemented and expanded for the particular requirements of a current problem, but this will prove to be a relatively demanding task. Alternatively, using SVMs on the whole (black box training + testing phase for each upcoming unknown sample) would slow down the effective application, especially if the number of known cases is large.

In this respect, the importance of the novel evolutionary kernel machine can be concretized through the following advantages:

- The ESVMs thus adopt the original learning strategy of the SVMs but aim to simplify and generalize the optimization component, by offering a transparent substitute by means of EAs to the Lagrange multipliers approach.
- Moreover, in order to converge, the evolutionary method does not require positive (semi-)definite kernels within nonlinear learning. Thus, any unconventional, but promising, kernel can be used.
- Since the coefficients are encoded in the structure of individuals, the equation of the evolving hyperplane is unrestrainedly available at all times and especially useful at the end of the process. The evolved mathematical combination of indicators may prove to be helpful to understand the weight and interaction of each attribute on the output. The accuracy of prediction of the model is computed using the available test samples, while the formula can be stored for future reference and employment when a new sample appears.

6.3.2 Formulation

The ESVM methodology therefore considers the adaptation of a flexible hyperplane to the given training data through the evolution of the optimal coefficients for its equation. Each individual encodes a candidate array of w and b and the whole set interacts towards the creation of enhanced solutions and the survival of the fittest against the training examples. After a number of generations, the EA converges to an optimal solution, which represents the best decision hyperplane that is both accurate and general enough. An intuitive representation of the process is shown in Fig. 6.1. The specific evolutionary steps to reach the optimal coefficients of a SVM decision surface are outlined in pseudo-code in Algorithm 6.1.

The next sections present in detail the specific components and mechanism of the involved EA.

6.3.3 Representation

An individual c is represented in (6.1) as an array of the coefficients of the hyperplane, w and b. Individuals are randomly generated, such that $w_i \in [-1,1], i = 1,2,...,n$ (recall that n is the number of features of a sample), and $b \in [-1,1]$.

$$c = (w_1, ..., w_n, b) \tag{6.1}$$

6.3.4 Fitness Evaluation

The fitness assignment derives from the objective function of the SVM optimization problem in (2.21) (Sect. 2.3.3 of Chap. 2) and is subject to minimization. Constraints

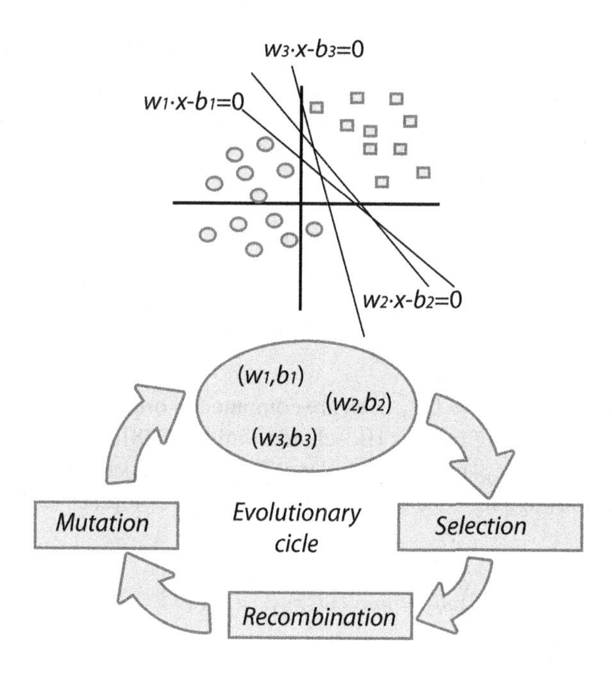

Fig. 6.1 The EA seeks for the coefficients of the decision hyperplane that optimally separates the classes, i.e. parameters (w, b)

Algorithm 6.1 The EA to establish the optimal equation of a SVM decision hyperplane.

Require: The optimization problem in (2.21) (Chap. 2)
Ensure: The fittest separating hyperplane with high training accuracy and generalization ability
begin
t ← 0;
Initialize population P(t) of potential hyperplanes of coefficients w and b;
Evaluate P(t) in terms of training accuracy and maximal separation between decision classes;
while termination condition is not satisfied **do**
  Parent selection;
  Recombination;
  Mutation;
  Evaluation;
  Survivor selection;
  t ← t + 1;
end while
return fittest hyperplane coefficients
end

---

are handled by penalizing the infeasible individuals through appointing a function $t : \mathbb{R} \to \mathbb{R}$ which returns the value of the argument, if negative, and 0 otherwise.

Consequently, the expression of the fitness function for determining the optimal coefficients of the decision hyperplane is defined as in (6.2).

$$
\begin{aligned}
f(w,b) = \|w\|^2 + C \sum_{i=1}^{m} \xi_i \\
+ \sum_{i=1}^{m} [t(y_i(w \cdot \mathbf{x}_i - b) - 1 + \xi_i)]^2
\end{aligned}
\tag{6.2}
$$

All error indicators ξ_i, $i = 1,2,...,m$, are computed in order to be referred into the evaluation. The procedure follows [Bosch and Smith, 1998]: the current individual (which is the present separating hyperplane) is taken and supporting hyperplanes are determined through the mechanism below. One first computes (6.3) :

$$
\begin{cases}
m_1 = min\{w \cdot x_i | y_i = 1\} \\
m_2 = max\{w \cdot x_i | y_i = -1\}
\end{cases}
\tag{6.3}
$$

Then one calculates (6.4):

$$
\begin{cases}
p = |m_1 - m_2| \\
w' = \frac{2}{p} w \\
b' = \frac{1}{p}(m_1 + m_2)
\end{cases}
\tag{6.4}
$$

For every training sample x_i, $i = 1,2,...,m$, the deviation to its corresponding supporting hyperplane is obtained through (6.5):

$$
\delta(x_i) = \begin{cases}
w' \cdot x_i - b' - 1, y_i = 1 \\
w' \cdot x_i - b' + 1, y_i = -1
\end{cases}
\tag{6.5}
$$

Subsequently, if the sign of the deviation equals that of the class, the corresponding $\xi_i = 0$. Else, the (normalized) absolute deviation is returned as an indicator for error. Normalization is necessary because the sum of deviations is added to the expression of the fitness assignment. As a consequence, in the early generations, when the generated coefficients lead to high deviations, their sum (considered from 1 to the number of training samples) takes over the whole fitness value and the evolutionary process is driven off the course to the optimum.

If nonlinear training is performed, the kernel may be either SVM-standard (restricted to obey Mercer's theorem for convergence), e.g. polynomial or radial, or may take any form whatsoever, as training is now performed under the flexible and unconstrained EAs.

6.3.5 Selection and Variation Operators

The efficient tournament selection and the common genetic operators for real encoding, i.e., intermediate recombination and mutation with normal perturbation, are applied (see Chap. 3 for details of these operators). Nevertheless, as before, different operators may be reasonable, as well.

6.3.6 Survivor Selection

The population of the next generation is decided to be formed of the newly obtained individuals plus the individuals that were not selected for reproduction. This means that the offspring resulting after either recombination or mutation automatically replace their parents. In the currently chosen case of recombination, where one offspring results from two parents, it is the less fit of them that is substituted by the descendant.

6.3.7 Stop Condition

The algorithm stops after a predefined number of generations and outputs the best hyperplane from the entire evolutionary process. As the fittest coefficients of the decision hyperplane, w^{opt} and b^{opt}, are found, the target for a new, unseen test data instance $\mathbf{x}'$ can be determined directly following the result of the function in (6.6).

$$f(\mathbf{x}) = w^{opt} \cdot \mathbf{x}' - b^{opt} \tag{6.6}$$

If the classification task is intrinsically binary, then the class is determined from the sign of the function as positive or negative. If the problem has several classes, then a voting mechanism is applied, based on the values of the current sample as a parameter of the different resulting decision functions [Hsu and Lin, 2004]. Finally, the classification accuracy is defined as the percent of correctly labeled cases with respect to the total number of test samples.

6.4 Experimental Results

Let us now evaluate whether the ESVM algorithm can produce competitive results when compared to the standard SVM approach. Four real-world test problems from the UCI Repository of Machine Learning Databases are used in our experiments. These are diabetes mellitus diagnosis, spam detection, iris recognition and soybean disease diagnosis. The motivation for our choices of test cases was manifold. Diabetes and spam are two-class problems, while soybean and iris are multi-class. Spam filtering data has a lot more features and samples than the diabetes diagnosis one, which makes a significant difference for classification as well as optimization. Conversely, while soybean has a high number of attributes, iris has only four, but a larger number of samples. The selected tasks thus contain various conditions for an objective validation of the ESVM approach.

Repeated random sub-sampling cross-validation is performed again for each data set. 30 runs of the ESVM are generated and, every time, approximately 2/3 random cases are assigned to the training set and the remaining 1/3 go into the test set. The values for all the parameters (of both the SVM and especially of the EA) are as well automatically tuned by SPO [Bartz-Beielstein, 2006]. The choice of SVM kernels is however the exception to this, where the linear and radial are tested in turn, as customary. For certain (e.g. radial, polynomial) kernels, the optimization problem shall be relatively simple, due to Mercer's theorem, and is implicitly solved by SVMs. Note that ESVMs are not restricted to using these traditional kernels, but we solely employ them to enable us to compare our algorithm with the classical SVMs. For the multi-class cases of iris and soybean, the commonly used one-against-one method [Hsu and Lin, 2004] is utilized.

For all test problems (except for soybean) the SPO indicates that recombination probabilities are dramatically increased, while mutation probabilities are often reduced. However, the relative quality of the final best configurations derived from SPO against the ones found during the initial LHS phase increases only with problem size.

We also performed a manual tuning of the variables, in order to test the ease of parametrization of the ESVM. It must be stated that, in most cases, results achieved with manually determined parameter values can be improved by SPO – if at all – only by increasing effort, like raising the population size or the number of generations. The parameters of the EA are therefore easily tunable, which is an important aspect when suggesting an alternative solution by means of such algorithms.

In order to compare the hardness of finding good parameters for the ESVM, the performance spectra of LHS are plotted for the spam and soybean problems in Fig. 6.2. The Y axis represents the fractions of all tried configurations. Therefore the Y value corresponding to each bar denotes the percentage of configurations that reached the accuracy marked on the X axis. The diagrams illustrate the fact that when SPO finds a configuration, it is already a promising one, as it can be concluded from the higher corresponding bars.

The hybridized ESVM construction produces equally good results when compared to the standard SVMs (see Table 6.1). However, the smaller standard deviations prove the higher stability of the ESVM approach. It must also be remarked that, for the standard kernels, one cannot expect ESVMs to be considerably better than the standard SVMs, since the kernel transformation that induces learning is the same. However, the flexibility of the EAs as optimization tools make ESVMs an attractive choice from the performance perspective, due to their prospective ability to additionally evolve problem-tailored kernels, regardless of whether they are positive (semi-)definite or not, which is impossible under SVMs.

The results for each problem were compared via a Wilcoxon rank-sum test. Computed p-values suggest to detect significant differences only in the cases of the soybean data set, where the ESVM is better.

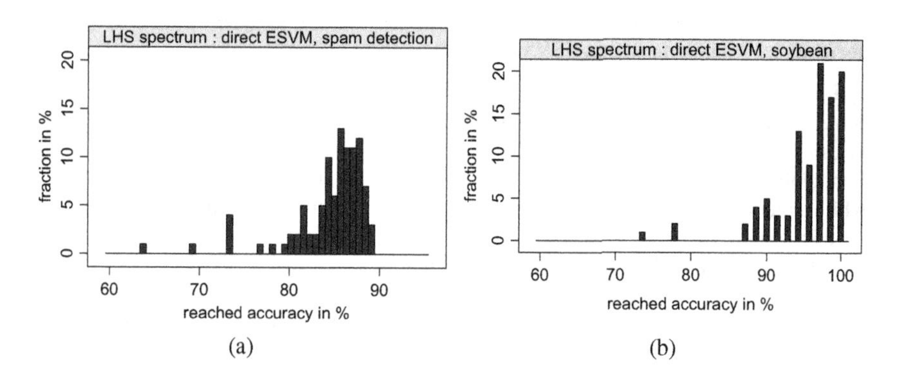

Fig. 6.2 EA parameter spectra of an LHS with size 100, 4 repeats, for ESVM on the spam (**a**) and soybean (**b**) problems [Stoean and Stoean, 2009b]

Table 6.1 Mean accuracies and standard deviations of standard SVMs on the considered test sets, in percent, as compared to those obtained by ESVMs

Data set	SVM	StD	ESVM	StD
Pima diabetes	76.82	1.84	77.31	2.45
Iris	95.33	3.16	96.45	2.36
Spam	92.67	0.64	91.04	0.80
Soybean	92.22	9.60	99.80	1.06

While observing the behavior of the ESVM on the benchmarking results, further possible applications came into attention, as well as inherent issues that eventually triggered enhancements within the proposed framework. These are all specified in the following subsections.

6.5 Dealing with Large Data Sets

A disadvantage of the methodology appears for large data sets where the amount of runtime needed for training is very large. This stems from the large number of necessary computations as the evaluation of each individual always refers all training samples. Consequently, this problem is tackled by an adaptation of a chunking procedure [Perez-Cruz et al, 2004] inside the ESVM [Stoean et al, 2009a].

The chunking mechanism is described by Algorithm 6.2. A chunk of N training samples is repeatedly considered. Within each chunking cycle, with a population of half random individuals and half previously best evolved individuals, the EA runs and determines the coefficients of the hyperplane. All training samples are tested against the obtained decision function and a new chunk is constructed based on half random incorrectly placed samples and half random samples from the current chunk. The chunking cycle stops when a predefined number of iterations passes with no improvement in training accuracy.

Algorithm 6.2 Chunking within ESVMs.

Require: The training samples and the optimization problem in (2.21)
Ensure: Fittest hyperplane coefficients and corresponding accuracy
begin
Randomly choose N training samples, equally distributed, to make a chunk;
while a predefined number of iterations passes with no improvement **do**
  if first chunk **then**
    Randomly initialize population of a new EA;
  else
    Use best evolved hyperplane coefficients to fill half of an EA population and randomly initialize another half;
  end if
  Apply EA and find coefficients of the hyperplane;
  Compute side of all samples in the training set with evolved hyperplane coefficients;
  From incorrectly placed, randomly choose (if exist) N/2 samples, equally distributed;
  Randomly choose the rest up to N from the current chunk and add all to new;
  if obtained training accuracy if higher than the best one obtained so far **then**
    Update best accuracy and hyperplane coefficients;
    Set improvement to true;
  end if
end while
Apply best obtained hyperplane coefficients on the test set and compute accuracy;
return accuracy
end

Experimentation was conducted on the classification problem of spam filtering. The employment of the chunking method made the training process run 8 times faster than without, which is already quite reasonable for the demanding real-world applications [Stoean et al, 2009a].

6.6 Feature Selection by Genetic Algorithms

SVMs have been demonstrated to circumvent the curse of dimensionality for highly multidimensional data sets [Joachims, 1998]. However, as we have previously seen in Chap. 5 as well, in practical complex tasks it may prove important to concentrate learning only on the most influential features from all given indicators. Taking the same example of a medical problem, the task may exhibit many medical factors [Stoean et al, 2011b], and even if ESVMs are able to give competitive results, they could also surely benefit from some form of prior feature selection. Moreover, as coefficients for each attribute are encoded into an individual, having many variables makes the genome too lengthy, which henceforth triggers more time for convergence. And, since EAs power the currently proposed technique, their flexibility can be further used in order to implement an ESVM embedded approach for choosing features, leading to improvement in its performance.

The methodology can be thus further endowed with a mechanism for dynamic feature selection provided by a GA (outlined in Algorithm 6.3). The GA-ESVM [Stoean et al, 2011b] is constituted of a population of individuals on two levels. The higher level of an individual selects optimal indicators, where the presence/absence of each feature is encoded by binary digits, while the lower one generates weights for the chosen attributes. Each level of an individual in the population is modified by its own genetic operators. The lower one is perturbed as before, while the higher GA parts of a chromosome undergo one-point recombination and strong mutation (see Chap. 3). Survivor selection is also performed differently at the high-level, since the offspring in the binary encoded component replace their parents counterpart only if better. The fitness assignment triggers evaluation at both levels, referring thus weights only for the selected features.

Algorithm 6.3 Selection of features and evolution of their weights by GA-ESVM.

Require: The optimization problem in (2.21)
Ensure: The fittest separating hyperplane with high training accuracy and generalization ability
begin
$t \leftarrow 0$;
Initialize population P(t) of potential hyperplanes of coefficients w and b, where each individual is made of two levels: a binary one where the powerful attributes are selected and a real-valued level encoding where weights for these chosen features are generated;
Evaluate P(t) in terms of training accuracy and maximal separation between decision classes, taking into account both the selected indicators and the weights in each individual
while termination condition is not satisfied **do**
  Parent selection;
  Recombination on the high-level population;
  Recombination on the low-level population;
  Mutation on the high-level population;
  Mutation on the low-level population;
  Evaluation;
  Survivor selection for the next generation;
  $t \leftarrow t + 1$;
end while
return fittest hyperplane coefficients along with its selected features
end

The enhanced mechanism was applied to the same problem of a correct discrimination between degrees of liver fibrosis in chronic hepatitis C [Stoean et al, 2011b]. The results did not only exhibit a statistically better learning accuracy, but also an indication of both the decisive factors within the assessment and their weights in the process [Stoean et al, 2011b]. Recall that the original data set contains 722 samples of 24 medical indicators and 5 possible degrees of fibrosis as output. For the differentiation of each class, the GA-ESVM dynamically selected those indicators that distinguish it from the other degrees. In this respect, the SVM one-against-all scheme is preferred, as it permits a more accurate discrimination between the particularities of each individual class and that of the rest. Starting from randomly chosen attributes, the algorithm cycle pursues an active evolution and an increase in the importance of certain features in the absence of others. Eventually, the coefficients for the remaining indicators represent weights for each given attribute. If the weight is closer to 0, then the corresponding feature has a smaller importance. If it conversely has a higher value, be that it is positive or negative, it will more powerfully influence the final outcome (as in Fig. 6.3) [Stoean et al, 2011b].

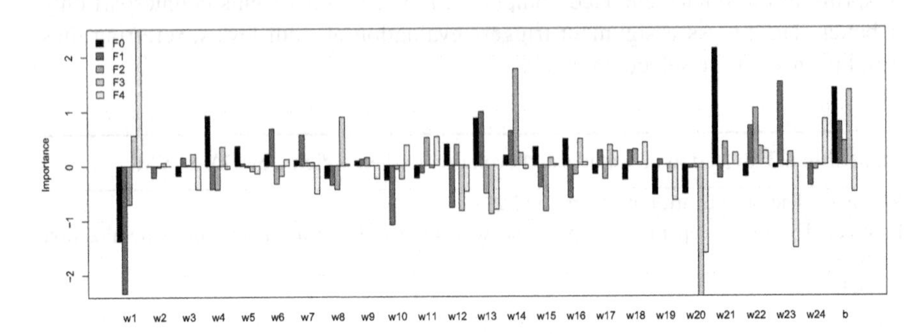

Fig. 6.3 The fittest coefficients determined by the GA-ESVM for every medical indicator (1-24) and for each of the 5 classes. Each group of 5 bars is related to one attribute. Longer bars (both positive and negative) correspond to more important attributes. Attributes omitted by the GA are represented with a weight of 0, i.e., w_2 for $F0$ and $F4$ [Stoean et al, 2011b]

6.7 Concluding Remarks

Several positive conclusions can be drawn from the feedback of the considered applications, whose results thus confirm to have met the goals of the ESVM architecture, planned as a viable alternative to SVMs for certain real-world situations:

- The ESVMs inherit the learning roots from SVMs and therefore remain efficient and robust.
- In contrast to the standard SVM solving, the optimization component now provided by means of EAs is transparent, adaptable and much easier to understand or use.
- ESVMs do not impose any kind of constraints or requirements regarding kernel choice. If necessary, for a certain more complex problem, it may even be simultaneously evolved, e.g. by a naïve approach like a HC that encodes a combination of different simple functions or by genetic programming [Howley and Madden, 2004].

- The ESVM is able to provide a formula of the relationship between indicators and matching outcomes. This is highly significant in order to obtain a rapid approximation of the response if certain values are appointed for the involved parameters, which may take place separately (from the training module) and instantly as a new example is given as input. This saves (computational) time when a critical case appears (e.g. in medicine).
- In addition, this formula is helpful for understanding interactions between variables and classes that may be difficult to grasp even for experienced practitioners. The weights also offer some insight on the importance of every attribute on the process.
- Moreover, the GA-ESVM is able to also dynamically determine the indicators that influence the result by an intrinsic feature selection mechanism.
- Most notably, the performance superiority of the alternative ESVM is statistically significant in comparison to the traditional SVM in some of the applications [Stoean et al, 2009a], [Stoean et al, 2009b], [Stoean et al, 2011b].

There are however also some possible limitations that may arise in the experimental stage:

- EAs make use of many parameters which are nevertheless relatively easy to be set in practical applications.
- Because of the reference to all samples at the evaluation of every individual, ESVMs have a slower training than SVMs, whose corresponding procedure makes only one call to the data set. However, in practice (often, but not always), it is the test reaction that is more important. Nevertheless, by observing the relationship between each result and the corresponding size of the training data, it is clear that SVM performs better than ESVM for larger problems; this is probably due to the fact that, in these cases, much more evolutionary effort would be necessary. Chunking partly resolves this issue, but other possibilities could be further investigated.

Chapter 7
Evolutionary Algorithms Explaining Support Vector Learning

All truths are easy to understand once they are discovered; the point is to discover them.
Galileo Galilei

7.1 Goals of This Chapter

Even if SVMs are one of the most reliable classifiers for real-world tasks when it comes to accurate prediction, their weak point still lies in the opacity behind their resulting discrimination [Huysmans et al, 2006]. As we have mentioned before, there are many available implementations that offer the possibility to also extract the coefficients of the decision hyperplane (SVM light, LIBSVM). In Chap. 6 we have also presented an easy and flexible alternative means to achieve that. Nevertheless, such output merely provides a weighted formula for the importance of each and every attribute. We have shown that feature selection can extract only those parameters that are actually determinant of the class and solve the issue of redundancy. However, the lack of any particular guidelines of the logic behind the decision making process still remains. This is obviously theoretically desired for a rigorous conceptual behavior, however it is also crucial for domains like medicine, where a good prediction accuracy alone is no longer sufficient for a true decision support for the medical act. While accuracy certainly remains a prerequisite [Belciug and El-Darzi, 2010], [Belciug and Gorunescu, 2013], [Gorunescu and Belciug, 2014] supplementary information on how a verdict had been reached, based on the given medical indicators, is necessary if the computational model is to be fully trusted as a second opinion.

On the other hand, classifiers that are able to derive prototypes of learning are transparent but cannot outperform kernel-based methodologies like the SVMs. The idea to combine two such opposites then sprung in the machine learning community: kernel techniques could bring the prediction force by simulating learning, while transparent classifiers could interpret their results in a comprehensible fashion.

There are many attempts in this sense, and several namely concerning the two subjects of this book: SVMs and EAs. In this context, the last chapter puts forward another novel approach built with the same target. We begin by addressing the existing literature entries (Sect. 7.2), present the new combination (Sect. 7.3) and enhancements (Sect. 7.5, 7.6 and 7.7), all from the experimental perspective (Sect. 7.4).

C. Stoean and R. Stoean, *Support Vector Machines and Evolutionary Algorithms for Classification*, Intelligent Systems Reference Library 69,
DOI: 10.1007/978-3-319-06941-8_7,

7.2 Support Vector Learning and Information Extraction Classifiers

A combination between a SVM and an explanatory classifier can be constructed on two grounds (see Fig. 7.1) [Martens et al, 2007]:

Pedagogical: SVMs establish a new input-output mapping, that is, each sample is labeled with the class predicted by the SVM. The relabeled samples are subsequently used by the information extractor. In other words, the SVM is actually a noise remover, which enables the following decision information extractor to concentrate learning only on correctly labeled data.

Decompositional: SVMs output the support vectors and the second method derives structured explanations from these. The support vectors are in fact the most important examples from the data set, as they shape the decision boundary. The approach also solves the runtime problem for the usual very large data sets connected to real-world problems. Therefore, this triggers both sample selection and noise removal prior to mining underlying rules.

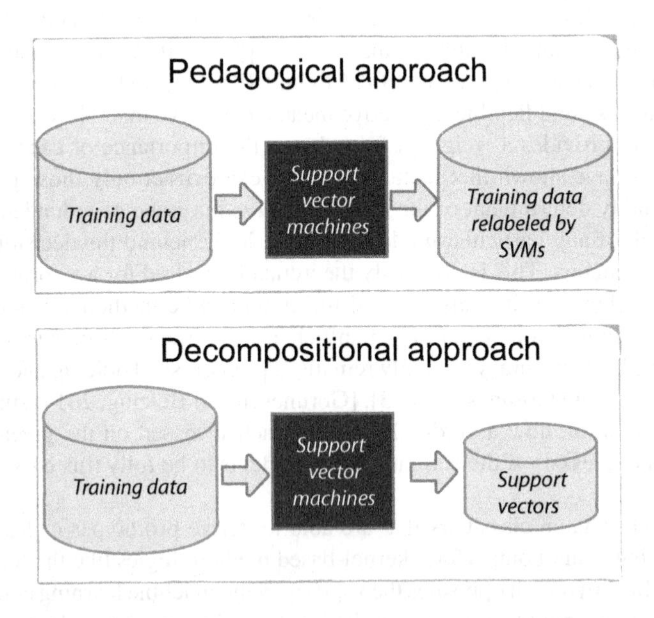

Fig. 7.1 Within the pedagogical approach, SVMs learn from the training set and then the classifier is applied to the same data to relabel it. The decompositional alternative simply extracts a small amount of the training data that represents the support vectors and these are kept with their original labels.

In order to meet its purpose, there are three consequential conditions that the final output of such a combined methodology must obey [Huysmans et al, 2006]:

Accuracy: The predicted targets for previously unseen samples must be more accurate than those derived from the information extractor alone.
Fidelity: Its results must approximate those of the black box SVM.
Comprehensibility: It must offer more comprehensible information than that of the initial learner, whichever form this knowledge may take.

We will next review the current such models that came to our attention. The first list involves different transparent knowledge extraction engines following a trained SVM model [Martens et al, 2007], [Diederich, 2008], [Farquad et al, 2010]:

- The SVM + Prototype approach of [Núñez et al, 2002] defines an ellipsoid through the combination of support vectors and data cluster prototypes to create an if-then decision scheme. The method suffers however from bad scalability.
- In [Fung et al, 2005], the problem is transformed to a simpler, equivalent variant and rules are constructed as hyper cubes by solving linear programs. This is not an advantage, as it can only be applied for linear decision kernels, which are generally not applicable for real-world data sets. Moreover, such an approach loses the strong ability of SVMs to model nonlinearities.
- In [Barakat and Diederich, 2005], the SVM relabeled input-output data are given to decision trees (DT) for the detection of the underlying learning system, while in the study [Barakat and Bradley, 2006] the area under the receiver operation characteristic curve is employed towards the same goal.
- In [Martens et al, 2009], an active learning-based approach is used to extract rules from support vectors.
- Finally, in [Farquad et al, 2010], the support vectors together with the actual output values of their targets are taken and provided to a fuzzy rule based system.

In the papers from the second list below, EAs are used in different formulations to collect the logic (mainly) behind neural networks (NNs) and SVMs. If we regard information extraction from the pedagogical point of view, then it makes no difference if we use SVMs, NNs [Haykin, 1999], [Gorunescu et al, 2011] or any other opaque classifier. That is the reason why we have included extraction from NNs in this list. A second motive is that, of all the combinations between opaque predictors and information extractors, hybridizations between SVM and EAs have been the least often explored.

- The GEX approach [Markowska-Kaczmar and Chumieja, 2004] learns from NNs, uses a special encoding for evolving rules and appoints an island model [Bessaou et al, 2000] to allow the existence of multiple subpopulations, each connected to a label of the problem to be solved. The disadvantage of this technique is that one sample can be covered by multiple rules, while it is not guaranteed that at least one rule will be valid for each class [Huysmans et al, 2006]. A changed EA, with a more elaborate representation for individuals and a Pareto multiobjective optimization behind, is provided later in [Markowska-Kaczmar and Wnuk-Lipinski, 2004].

- The G-REX alternative [Johansson et al, 2010] is a more accurate general technique that uses genetic programming [Langdon and Poli, 2001] to extract rules of various representations from different (opaque or not) models (NNs, random forests). The approach in [Martens et al, 2007] applies the G-REX method to SVMs instead of NNs.
- The methodology in [Ozbakir et al, 2009] achieves a combination between NNs and ant colony optimization [Dorigo and Stützle, 2004], [Pintea, 2014] for the same task.

7.3 Extracting Class Prototypes from Support Vector Machines by Cooperative Coevolution

Within these premises, we can formulate a novel combined method by appointing the CC algorithm (in Chap. 5) to discover the class prototypes after the data set had been processed by the SVM.

7.3.1 Formulation

The construction of the hybridized method under discussion [Stoean and Stoean, 2013a], [Stoean and Stoean, 2013b] is intuitively illustrated in Fig. 7.2. The flow of the technique can be therefore formulated in short as follows:

1. The SVM reshapes the data
    a. either in a pedagogical fashion
    b. or in a decompositional way.
2. The CC is trained on these changed data sets and determines attribute thresholds (as described before in Chap. 5).
3. The resulting attribute thresholding for each class must be
    a. accurate to new samples,
    b. faithful to the opaque model,
    c. as simple and compact as possible, since an intricate and hard to follow architecture may actually offer less comprehensibility.

7.3.2 Scope and Relevance

There are several advantages arising from this new approach for extracting the prediction hidden observations of SVM (as compared to related attempts in the literature):

- EA individuals can directly encode thresholds for problem indicators. Comprehensibility can thus be successfully achieved by generating prototypes for each class of the task, while they are also easily maneuvered by the EA.

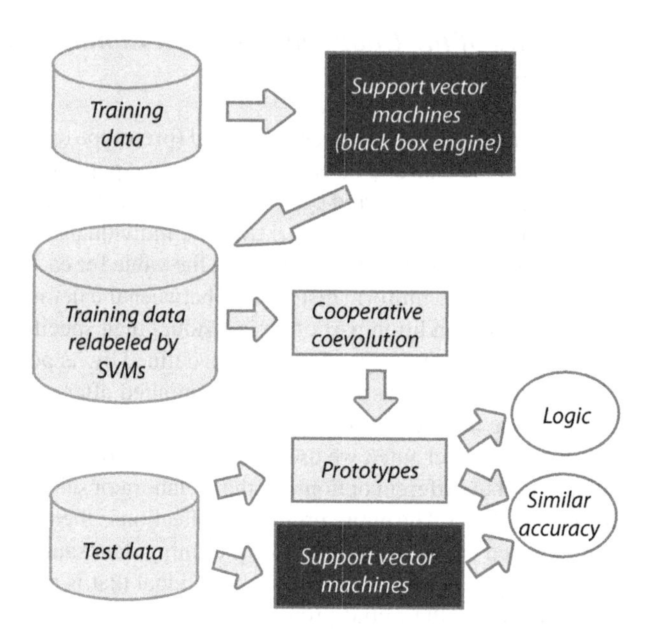

Fig. 7.2 SVMs are used to *clean* the given samples, CC is then trained on the new data set and creates prototypes that are used to better classify the test data. Additionally, CC provides explanations via the prototypes for each class as regards the logic behind the decision making process.

- The IF-THEN format holding conjunctive statements with equal signs for referring thresholds is also simpler to follow. Several inequalities or a complex format (like in [Markowska-Kaczmar and Chumieja, 2004], [Markowska-Kaczmar and Wnuk-Lipinski, 2004] or [Johansson et al, 2010]) cannot but harden the reading of the decision explanations.
- The extraction CC engine, as an EA, is an adaptable framework to implement different possibilities of resolving the task, as previously seen in its application results for classification in Chap. 5.
- As concerns the diversity of resulting class prototypes, CC inherently maintains several distinct concurrent subpopulations, as the number of classes of the problem determines the number of species. This multimodal mechanism is thus more straightforward when evolving distinct prototypes for the different classes of the decision problem (unlike the island model in [Markowska-Kaczmar and Chumieja, 2004]), as each class triggers one population. Thus, prototypes of every class eventually become homogenous, but they remain very different from those of the other species. This is also another reason why CC was preferred as the multimodal engine, instead of the alternative GC (in Chap. 4).
- The encoding is thus even simpler than the usual rule formation, since the class is not part of the prototype, resulting directly from the subpopulation it is connected to.

7.3.3 *Particularities of the Cooperative Coevolutionary Classifier for Information Extraction*

As before in the CC approach (Chap. 5), each individual (prototype or rule) encodes values for all indicators in the data and its class is given by the population it belongs to. Hence, its formal expression is referred again as (4.3), where the prototype is representative of class y_i of the problem, $i = 1, 2, ..., k$. The individuals of the starting population are once more randomly initialized, where the value for each of the n attributes is generated following a uniform distribution between the definition bounds of that specific feature. The condition part of an individual then specifies indicator thresholds that designate it as a prototype for the class defined by its population.

The prediction capability of a class prototype is computed after a complete set is formed by selecting one individual from each of the other subpopulations. In the experiments that follow in this chapter, we use a random selection of the individuals from the subpopulations, but different options for the collaborator selection pressure parameter could be considered. The entire prototype collection is then applied to the training data as remodeled by the SVM. For every training data sample, distances to each collected prototype are calculated and the individual that is closest decides its label. The performance of the initial prototype is then given by the prediction accuracy over all training samples.

While comprehensibility is thus primarily resolved through the EA individual representation, the two requirements regarding fidelity to the SVM and high prediction accuracy are met through reference through the CC fitness expression described in the lines above. The actual place of inclusion is when success is measured by comparing the outcome of a sample with the SVM-CC prediction. If the approach is pedagogical, then the actual outcomes for the training data examples are those confirmed by the SVMs. Fidelity is thus addressed as in (7.1) and expresses the percentage of identically labeled samples [Huysmans et al, 2006]. x_i is a sample, y_i^{SVM} is its outcome as predicted by the SVM and y_i^{SVM-CC} that which is provided by SVM-CC, $i = 1, 2, ..., m$.

$$fidelity^{SVM-CC} = Prob(y_i^{SVM} = y_i^{SVM-CC} | x_i \in [a_1, b_1] \times [a_2, b_2] \times ... \times [a_n, b_n]) \tag{7.1}$$

If the behavior is decompositional, the real outcomes of the support vectors are those given in the initial training data set. Accuracy [Huysmans et al, 2006] is therefore also obeyed as in (7.2).

$$accuracy^{SVM-CC} = Prob(y_i^{real} = y_i^{SVM-CC} | x_i \in [a_1, b_1] \times [a_2, b_2] \times ... \times [a_n, b_n]) \tag{7.2}$$

Finally, as regards the test stage, the corresponding samples are classified by a set of prototypes appointed from the final subpopulations and their predicted outcomes are confronted with those present in the original data set.

The approach is sketched by Algorithm 7.1.

Algorithm 7.1 CC for extracting learning prototypes from SVMs.

```
Require: A k-class classification problem
Ensure: A rule set with multiple prototypes for each class
  begin
  if approach is pedagogical then
    Relabel labels of training data as predicted by the SVM;
  else
    Collect the support vectors with their actual labels from the data;
  end if
  t ← 0;
  for each species i do
    Randomly initialize population P_i(t);
  end for
  for each species i do
    if approach is pedagogical then
      Evaluate P_i(t) by selecting collaborators from the other species for every individual
      and compare classes according to fidelity;
    else
      Evaluate P_i(t) by selecting collaborators from the other species for every individual
      and compare classes according to accuracy;
    end if
  end for
  while termination condition is not satisfied do
    for each species i do
      Select parents from P_i(t);
      Apply genetic operators;
      if approach is pedagogical then
        Evaluate P_i(t) by selecting collaborators from the other species for every individual
        and compare classes according to fidelity;
      else
        Evaluate P_i(t) by selecting collaborators from the other species for every individual
        and compare classes according to accuracy;
      end if
      Select survivors from P_i(t) to P_i(t + 1);
    end for
    t ← t + 1;
  end while
  return a complete set of prototypes for each class
  end
```

7.4 Experimental Results

We want to assess three goals in order to prove the effectiveness of the proposed combined SVM-CC approach for white box extraction:

- fidelity to SVMs;
- accuracy superior to CC;
- comprehensibility superior to SVMs.

The first two aims test the viability of the hybridization, such that the SVM-CC performs better than the CC and comparable to the SVM. Thirdly, SVM-CC must also provide the class prototypes underlying the decision model in a form that must be understandable for the reader. Only after having those validated, we can safely affirm that the methodology accomplishes the theoretical goal of such a combination of classifiers and offers credible practical assistance.

The choice for the test problems comes yet again from the UCI repository and we target breast cancer, diabetes mellitus and iris discrimination.

Once the SVM pedagogical and decompositional steps were over and the training data was relabeled, we plotted the average number of samples in the training set that have the outcomes changed by the SVMs (pedagogical), besides the average number of support vectors (decompositional), both against the total number of training samples for each data set (see Fig. 7.3). Note that the sizes of the bars should be compared only to the ones within the same group. The number of support vectors is higher than the number of samples with changed outcomes, but significantly lower than the cardinal of the complete training data. This observation implies that a major reduction of dimensionality (in the number of samples) is performed prior to the application of CC in the decompositional case. Although for the iris data set the samples whose outcomes are changed (first bar) seem almost absent in the figure, the average value is in fact of 1.87.

The experimental setup is established as follows. As variation operators, we once more choose those common for a continuous encoding, i.e. intermediate recombination and mutation with normal perturbation. The binary tournament type is taken again as the selection operator. The size of each subpopulation is set to 50; it is however always taken k times the value (k being the number of classes), because there exists one population connected to each class. The number of evolutionary loops is considered as 80. The values for the mutation strength, mutation and recombination probabilities are chosen using the SPO [Bartz-Beielstein, 2006]. The values for the probabilities are picked from the [0, 1] interval, while for the mutation strength they are taken from [0, 2]. Each data set is again 30 times randomly split into 2/3 training and 1/3 test samples. The 30 training/test sets are the same in all approaches, for the SVM preprocessing to be identical. The reported prediction accuracy is obtained by averaging, over the 30 different runs, the percent of correctly labeled samples from the test set out of their total. When computing the fidelity to the SVMs, the known labels of the test samples are those given by the SVM output, then a typical accuracy ratio is computed once more.

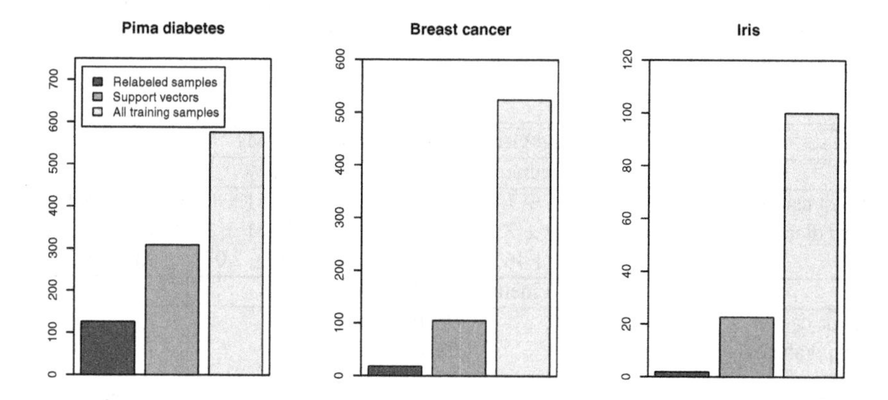

Fig. 7.3 Number of samples (on the vertical axis) changed by the SVMs within the pedagogical methodology, besides the number of support vectors from the decompositional approach and the total number of training samples for each data set

Table 7.1 outlines the prediction accuracies on the corresponding data sets obtained by CC alone, the pedagogical and decompositional SVM-CC, the SVMs and DT [Gorunescu, 2011]. The last methodology is included in the experiments, in order to investigate whether the SVM-CC white box extraction method performs better than a one-step transparent DT model applied directly to the initial data set [Stoean and Stoean, 2013a]. Additionally, we want to test if it may serve as an alternative to the CC as the explanatory engine. The p-values computed via a Wilcoxon rank-sum test show significant differences in results for diabetes (for the SVMs over CC, decompositional SVM-CC and DT) and iris (for both CC and SVM-CC variants over SVMs and DT).

In the last line of Table 7.1, we can also see that the fidelity criterion is obeyed by the hybridized approaches. High fidelity is very much desired, since it implies that the combined approach capably uses the SVM relabeling of the training data to learn the relationship between the values for the attributes and the triggered outcomes. It then uses the information to classify previously unseen data (almost) as efficiently as the SVMs. Fidelity is computed by measuring the similarity of test prediction between the combined approach and the SVM. Good fidelity however also conducts to better prediction accuracy, since SVM-CC usually behaves very similarly to SVMs, as observed in the fidelity outcomes, and SVMs represent a great choice of a classification algorithm to mimic.

If we look at the fidelity results in comparison to the prediction outcomes in Table 7.1, the relabeled samples from the training set clearly represent more accurate data for the SVM-CC approaches, since the fidelity values are with no doubt higher in general than the corresponding accuracies. This proves that SVM labeling eliminates noise from the data and learning becomes more efficient.

Table 7.1 Comparison between prediction accuracy results and standard deviations obtained on the test data sets by the considered approaches averaged over 30 repeated runs. The last rows additionally show fidelity to SVM predictions.

Data set	CC	Pedagogical	Decompositional	SVMs	DT
		Average accuracy $\pm$ standard deviation (%)			
Breast cancer	96.78 $\pm$ 1.21	96.95 $\pm$ 1.14	95.92 $\pm$ 1.26	96.51$\pm$ 1.41	94.11 $\pm$ 1.61
Pima diabetes	75.12 $\pm$ 3.63	76.49 $\pm$ 3.31	72.26 $\pm$ 3.85	77.31 $\pm$ 3.37	74.31 $\pm$ 2.92
Iris	97.6 $\pm$ 1.58	98 $\pm$ 1.46	98.07 $\pm$ 1.21	96 $\pm$ 2.07	93.73 $\pm$ 2.45
		Fidelity to SVMs (%)			
Breast cancer	-	98.02	96.46	-	-
Pima diabetes	-	90.71	78.7	-	-
Iris	-	96.93	97.2	-	-

The pedagogical approach appears to be more consistent in accuracy results than the decompositional one. Also, it is more faithful to the SVM outcome. Its predictions from Table 7.1 can also be seen as better than the ones of the CC and closer to those of the SVMs. However, there is a great enhancement in runtime for the SVM-CC decompositional approach, since the training data set is drastically reduced to solely the support vectors.

When comparing the DT results with the ones of the CC alone, we can see that there is a small advantage for the latter: statistical testing confirms that CC is significantly better for the 2 disease diagnosis problems and equal for iris. The direct comparison between CC and DT was performed not only as to check whether the latter could serve as a better alternative for the explanatory algorithm in the two-step approaches, but also to underline the need for such combined techniques. The DT results are far below the ones of the proposed SVM-CC for the 3 data sets, so they cannot be a viable replacement.

Looking simultaneously at Fig. 7.3 and Table 6.1, it can be noticed that there is a strong relation between the average test accuracy results of the SVMs and the number of training samples that they relabeled. This number, plotted as the first bar in each group should be assessed as opposed to the third bar of the group that stands for the total number of training samples. For breast cancer and iris, where test accuracy goes beyond 90%, the number of labels changed by SVMs in the training set is very small. The reason is that such a data set already has the samples of different classes well separated, the noise amount in the data is low, so there are only a few problematic samples from the point of view of the SVMs. A similar correlation can also be observed between a low number of support vectors and the success of the SVMs again for iris and breast cancer.

For an intuitive understanding of the class prototypes, the discovered thresholds for attributes of each the three data sets are plotted in a random run of the pedagogical approach (see Fig. 7.4). One prototype is connected to each class and the latter is designated by a specific symbol. A class prototype is read by following the lines with a certain sign from the first attribute to the last one. The exact discovered thresholds can be observed in the plots from the left hand side, but not all features have the same domain, so it is not relevant to compare the vertical distances for pairs of attributes. For a proper comparison between the thresholds, the same obtained values are normalized to the [0, 1] interval in the plots on the right hand side. Such a visualization helps to understand where are the thresholds situated on the initial intervals (plots on the left column) and what are the critical differences between attribute values discovered for the different classes (plots on the right column). The breast cancer problem however has each of the nine attributes defined on the same [0, 10] interval, so there is no difference between the positioning of the thresholds on the plots.

We can thus clearly see how some attributes count more than others. For the iris data set, for instance, it is the third attribute that makes a clear difference, while the thresholds for the others are very similar for all outcomes. For the Pima diabetes case, it is normalization that helps in distinguishing the importance of several features. While in the left plot there are many attributes that appear to have threshold values near one another and visually look alike, on the right one we can see that they are actually not that close - see attributes 1, 4, 6-8. The available class prototype set and its picture can prove helpful in supporting practical decision making, since the user can get usually fast aware of both the thresholds of demarcation between the classes and also of the relevance of each problem feature for the task.

7.5 Feature Selection by Hill Climbing – Revisited

When dealing with a large number of indicators, like those that define data in the medical field, where many of the attributes have little discriminative power between the potential outcomes, a means to reduce their number is especially important. The presence of too many attributes can divert classifiers as well as physicians from distinguishing those whose values differentiate between diagnoses. As also previously discussed, feature selection has been shown to help towards a faster and more accurate classification [Akay, 2009]. This commonly takes place before the actual classification, however, it can also be included as part of a cycle inside the classifier, which learns with differently selected features until some condition is met.

We were confronted with this problem while running the experiments for a first study on breast cancer diagnosis and generation of decision explanations by SVM-CC [Stoean and Stoean, 2013b]. We have said before that it had been shown in [Joachims, 1998] that the inner workings of SVMs bypass the dimensionality issue. The CC classifier cannot avoid it, however the adjustability of an EA framework offers the possibility to embed a feature selector within the evolution of classification thresholds [Stoean et al, 2011a], like we have seen before in Chap. 5. Therefore, once again, a dynamic chemistry between chosen indicators and their proper thresholds is performed. This interaction changes thresholds for attributes, as for every new HC individual different dependencies are involved and the SVM-CC is re-initiated with each change in the HC configuration. The reference to fewer indicators additionally offers more comprehensibility to the generated prototype set.

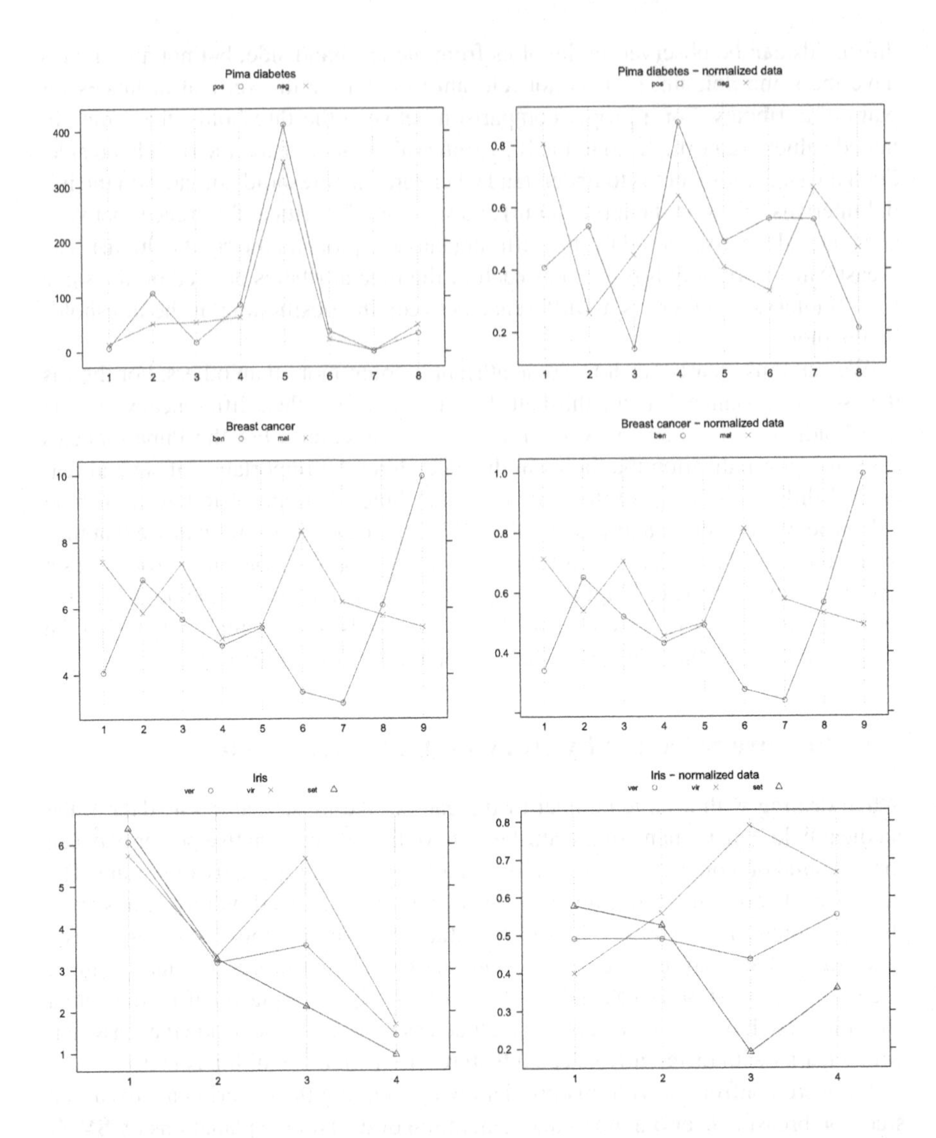

Fig. 7.4 The discovered class prototypes for the three data sets in a random run of the SVM-CC pedagogical approach. First column plots the obtained thresholds for the raw data, while within the second the values are normalized to the interval [0, 1]. The horizontal axis refers the attributes by their number and the vertical one shows their values. The different signs stand for the distinct classes of the problems (excerpt for [Stoean and Stoean, 2013a]).

Since it exhibited better results in the initial experimentation, it was the pedagogical approach that we selected for attaching the feature selector. The experimental setup is changed from the version in Chap. 5, as we can now also refer a goal accuracy value for each problem (the one attained by the SVM). The HC runs until it reaches this value (best case scenario) or, if no improvement is achieved for 50 iterations, it is re-initialized and re-run. This re-initialization may happen for up to 5 times and if the targeted percent is not reached, it is the current accuracy that is returned. This desired goal is set for trying to break the limits of the SVM-CC algorithm as concerns its obtained average accuracy. The evaluation of the HC individual presumes a typical run of the approach and the obtained accuracy represents the fitness outcome. The HC algorithm conducts an iterated search for picking the most appropriate combination between the attributes of the classification problem and the weights discovered by the SVM-CC method for those features.

The prediction accuracy for the breast cancer problem now reached 97.16%, which is not significantly better than without the HC. The number of attributes is nevertheless reduced in average from 9 to 5. This means not only that noisy information is removed, but the user can also more easily grasp the decision prototypes, and consequently the classification problem. The most important features, i.e., those that are included into several prototypes (subsets of attributes) are thus evidenced. All these nevertheless bring an accompanying longer runtime, as the HC calls the pedagogical method at each iteration.

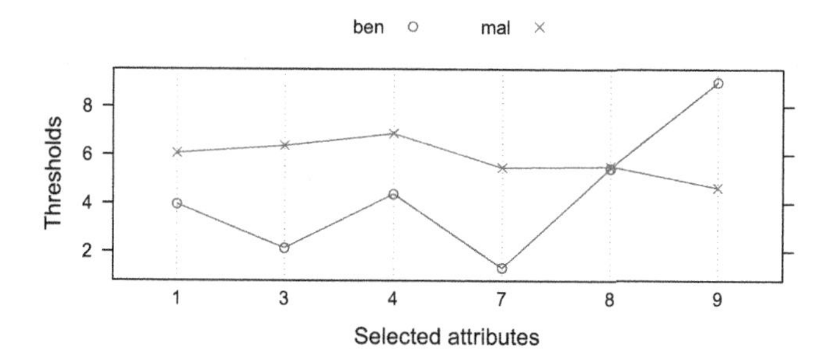

Fig. 7.5 The selected attributes and their generated thresholds after a random run of the HC as feature selector on the breast cancer diagnosis problem (excerpt from [Stoean and Stoean, 2013b])

The HC selected features and their corresponding thresholds are outlined in Fig. 7.5. As previously in Fig. 7.4, the prototypes from Fig. 7.5 provide an intuitive description for the discovered information. Larger distances on the vertical axis mean that there is a greater disparity for the values of that attribute and the indicator makes a clearer difference for the current prototype as opposed to attributes that have the values closer. In fact, it may happen that, for a different training configuration, the thresholds of the same attribute shall be close to each other. The indicator thresholds should hence not be read out of the context, but they should only be analyzed together with the configuration of the entire prototype. Recall that the prototypes are not unique, but they emphasize the connection between the values of different variables.

As before in Chap. 5, we also measure which attributes are more often selected. The results show that all indicators participate to the construction of the prototypes, some more often and some only seldom (Fig. 7.6). This means that good configurations of attribute thresholds can be achieved for different feature subsets, but surely some of them are more important since the best results are obtained when they are always part of the selection.

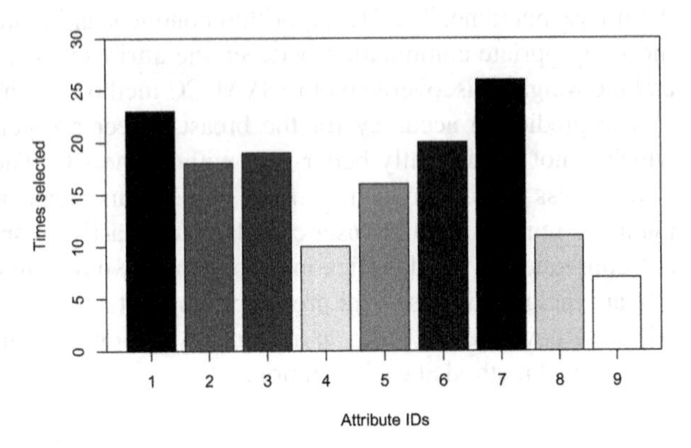

Fig. 7.6 Most often selected attributes (out of 30 HC runs) in breast cancer diagnosis (excerpt from [Stoean and Stoean, 2013b])

The attribute that was selected most by the HC (i.e., attribute 7, as seen in Fig. 7.6) is put to alone discriminate the data between the two classes of breast cancer diagnosis (Fig. 7.7). Based on the values of that specific attribute (on the horizontal axis), the samples are distributed at a high degree (on the vertical axis) between the two classes.

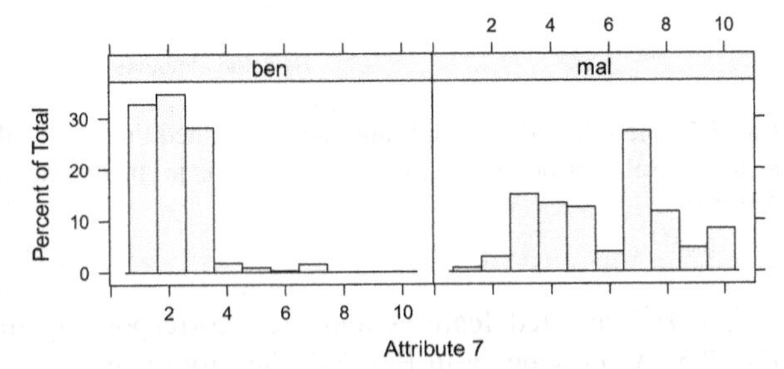

Fig. 7.7 Breast cancer sample distribution between the two outputs (benign and malignant) on the basis of the most important attribute as observed in Fig. 7.6 (excerpt from [Stoean and Stoean, 2013b])

7.6 Explaining Singular Predictions

Reducing the number of attributes by feature selection surely resolves the intricacy of the problem. But a medical expert would still have doubts as concerns the output of the model. A set of prototypes that is computationally easy to apply to a large set of test samples might not be very comprehensible for the user when it should classify a single record. The physician would surely prefer an indication towards the features that determined the classifier to put a certain diagnostic for a patient [Strumbelj et al, 2010].

To also accomplish this with our approach, after a sample is classified, the absolute differences between the values of that record and the corresponding value of the prototype for that found class can be calculated for each attribute. By ascendingly ordering the attributes according to the values obtained for these absolute differences, a measure of their relevance (first being the most relevant) for the taken decision is achieved. Even if this methodology is extremely simple [Stoean and Stoean, 2013b], it outlines the individual attributes that have the closest values to the weights (as given by the prototypes) of those indicators which determine the diagnosis of a certain class.

The following experiment is further on performed for the breast cancer data: the first two patients with different diagnoses are chosen and the differences mentioned before are computed. Gathering the most important attributes for the first 10 patients taken under equally balanced diagnoses, the order for the most decisive attributes is found 1 and 7 for patients diagnosed as malign and the same attributes but in reverse order for the benign individuals. If we look again at Fig. 7.6, the former experiment revealed the importance of the same two attributes.

Such computations prove very useful in practice because they can be obtained for a current patient individually and they point out which are the most relevant indicators of the diagnosis in that case. By taking only small amounts of data from each problem, some of the features that were previously found as decisive by the more computationally expensive HC (in Fig. 7.5 and 7.6) are confirmed through this simple individual oriented classification.

7.7 Post-Feature Selection for Prototypes

After the evolution of prototypes ends, a set of resulting distinct solutions is selected (i.e., a collaborator from each class) to be tested against samples from the test set. If the decision set is inspected, it is natural that the thresholds for certain attributes may be closer to each other than others in different prototypes. We assumed that these attributes have little or no influence as concerns the classification of a new sample.

Algorithm 7.2 outlines a posterior feature selection to eliminate from each solution the attributes whose thresholds are very close to a mean over all prototypes for the corresponding values [Stoean and Stoean, 2013a]. Besides the decision set, the algorithm receives a positive integer, which is the significance threshold s under which less significant features are discarded. The values for s start from 0, when no attribute is removed, and can be incremented until a prototype remains with no attributes. Actually, the value of parameter s represents a percent of the definition span of the current attribute.

The algorithm begins by creating a vector of mean values, whose size is equal to the number of attributes of the classification problem. The value of locus i represents the average over all thresholds on position i of the considered prototypes. Then, for each rule and accordingly for each class, we find the attribute that is most distant with respect to the vector of means. This similarity is normalized for each attribute in order to have a relative comparison. Such a value is important in determining to what extent can the significance threshold be increased until a prototype is completely eliminated because all its attributes are marked as unimportant. The fact that each solution has at least one attribute with a significant value is assured in the condition line that verifies if $(b_i - a_i) \cdot s/100 < threshold^l_{dist}$. Subsequently, each attribute of every rule is considered and a difference in absolute value is computed against the rates from the vector of means. If the obtained positive number is lower than the significance threshold, then this attribute is ignored for the current prototype. Note that for classification problems with more than two classes, an attribute may be removed from a prototype, but it is further kept in a complementary one, as it can be important for one class, but insignificant for others.

Algorithm 7.2 Post-feature selection for class prototypes.

Require: The set of k prototypes, k being the number of classes, and a significance threshold s for attribute elimination
Ensure: The k prototypes holding only the relevant attributes
  begin
  Compute vector *mean* of length n by averaging the values for each attribute threshold over all the prototypes $\{n$ is the number of attributes$\}$
  for each prototype l **do**
    Find $threshold^l_{dist}$ among all $threshold_i$, where $i \in \{1, 2, ..., n\}$, that is the remotest to $mean_i$, i.e., corresponds to $\max_{i=1}^{n} \frac{| threshold_i - mean_i |}{b_i - a_i}$
  end for
  for each prototype l **do**
    if $(b_i - a_i) \cdot s/100 < threshold^l_{dist}$ **then**
      for each $attribute_i$ **do**
        if $|threshold_i$ - $mean_i| < (b_i - a_i) \cdot s/100$ **then**
          Mark i as a *don't care* attribute for prototype p
        end if
      end for
    end if
  end for
  end

What remains to be reformulated is the corresponding change in the application of prototypes of different lengths to the test set. The distance from an unknown sample is now applied only to the attributes that matter from the current prototype and it is divided by the number of solely these contributing features. The motivation for this division lies in the fact that some prototypes may have many relevant attributes, others can remain with a very low number and in this way the proportionality of comparison still holds.

We chose again the more stable pedagogical approach for testing this planned enhancement and we tried values for the significance threshold to the maximum possible number. We applied it only to Pima diabetes and breast cancer diagnosis, as the iris data already has only 4 attributes. The obtained accuracy, as well as the number of eliminated attributes for the two data sets, are shown in Fig. 7.8. The horizontal axis contains the values for the significance threshold. On the vertical axis there is one line with accuracies followed by another with the number of removed attributes, in order to have a simultaneous comparison. A change in accuracy is almost absent. The gain is nevertheless represented by the fact that the number of dimensions is substantially reduced and the remaining thresholds for the decisive attributes within the class prototypes can be more easily analyzed and compared.

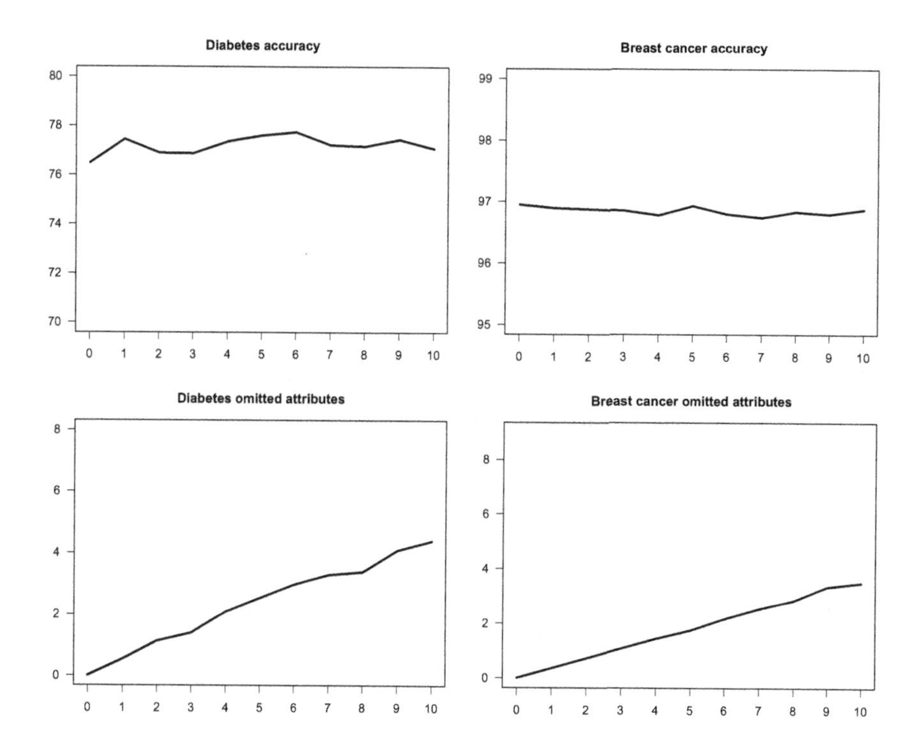

Fig. 7.8 The plots on the first line output the accuracies obtained for Pima diabetes and breast cancer diagnosis, when attributes are eliminated. The graphics from the second line show how many features are discarded for each. The horizontal axis contains values for the significance threshold used in eliminating the attributes, while accuracy (line 1) and number of removed attributes (line 2) are represented on the vertical axis (excerpt from [Stoean and Stoean, 2013a]).

Figure 7.9 plots the new prototypes of every class for the largest significance threshold taken for each case in Fig. 7.8, that is the highest value on the horizontal axis. Since the classification problems targeted here have only two classes, an attribute is eliminated from both prototypes at the same time. If the task is however multi-class, an attribute may be discarded only from certain prototypes. This happens because attribute elimination is applied by making use of an average over the prototypes for all classes. In the binary case, the thresholds for the two prototypes have an equal distance to the mean, so they are both or none eliminated [Stoean and Stoean, 2013a].

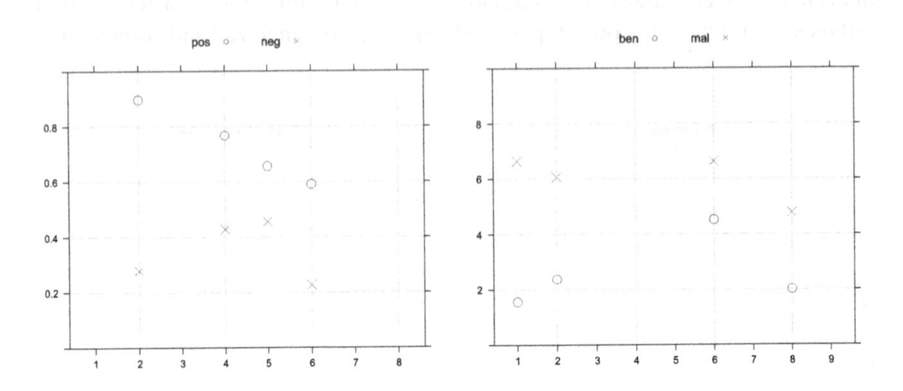

Fig. 7.9 Illustration of the remaining attributes and their threshold values for the highest significance threshold following Fig. 7.8 - Pima diabetes data left and breast cancer diagnosis on the right (excerpt from [Stoean and Stoean, 2013a])

7.8 Concluding Remarks

Following all the earlier experiments and observations, we can draw the following conclusions as to why is SVM-CC – a combination between a kernel-based methodology and a prototype-centered classifier – a good option for white box extraction:

- The EA encoding is simpler than genetic programming rules [Johansson et al, 2010] and the separation into explicit multiple subpopulations, each holding prototypes of one class, is more direct than ant colony optimization [Ozbakir et al, 2009] and easier to control than island models [Markowska-Kaczmar and Chumieja, 2004] or genetic cromodynamics [Stoean et al, 2007], [Stoean and Stoean, 2009a].
- The possibility of easily including a HC into the representation triggers simultaneous feature selection and information extraction.
- The option of allowing only the presence of the informative indicators additionally facilitates a deeper and more pointing understanding of the problem and relevant features, all centered on their discriminative thresholds.

- The derived prototypes discover connections between various values of different attributes.
- Through the individual classification explanation, the expert is able to see a hierarchy of the attributes importance in the automated diagnosis process.
- A second method of reducing the complexity of the decision guidelines by omitting for each class the attributes that had very low influence for the corresponding discrimination prototype leads to a compact formulation of the involved attributes. This helps discover relevant insights on the problem at hand, as well as shows a more understandable picture of the underlying decision making.
- When feature selection is performed, either online or a posteriori, the accuracy does not increase, but comprehensibility nevertheless does grow. This is somehow obvious, since the less informative attributes were probably weighted less and had small or no influence on the resulting predictions.
- If we were however to compare the comprehensibility power of both feature selection approaches, we would say that cleaning each prototype of the insignificant indicators for that outcome leads to better understandability than the online approach that performs global feature selection on the data set. This is due to the fact that the a posteriori method reveals the specific interplay between selected attributes for each particular class in turn.

Chapter 8
Final Remarks

Begin at the beginning, the King said, very gravely, and go on till you come to the end: then stop.
Alice in Wonderland by Lewis Carroll

We have reached the end of the book. Have we got any answers to the question we raised in the introductory chapter? Having presented the several options for classification – SVMs alone, single EAs and hybridization at two stages of learning – what choice proved to be more advantageous, taking into consideration prediction accuracy, comprehensibility, simplicity, flexibility and runtime?

Let us summarize the conclusions we can draw after experimenting with each technique:

- SVMs (Chap. 2) hold the key to high accuracy and good runtime. If good accuracy and small runtime are all that matter for the user, SVMs are the best choice. If interest also concerns an intuitive idea of how a prototype for a class looks like, the user should see other options too, as SVMs put forward insufficient explanations [Huysmans et al, 2006] for how the predicted outcome was reached. That which cannot be understood, cannot be fully trusted.
- GC (Chap. 4) and CC (Chap. 5) evolve class prototypes holding attribute thresholds that must be simultaneously reached in order to label a sample with a certain outcome. Evaluation requires good accuracy on the training data. Diversity for prototypes of distinct classes is preserved either through radii separating subpopulations or through the maintenance of multiple populations, each connected to one class. Understandability is increased, as thresholds are provided for the attributes that differentiate between outcomes, but accuracy cannot surpass that of SVMs. Feature selection can be directly added to the evolutionary process through a concurrent HC. Not all problem variables are thus further encoded into the structure of an individual, which leads to genome length reduction and enhancement of comprehensibility.
- These findings trigger the idea of combining the good accuracy of SVMs with the comprehensible class prototypes of EAs. ESVMs (Chap. 6) geometrically discriminate between training samples as SVMs, formulate the SVM primal optimization problem and solve it by evolving hyperplane coefficients through EAs. This is in fact EAs doing the optimization inside SVMs. Accuracy is comparable to SVMs, the optimization engine is simpler and the kernel choice unconstrained.

C. Stoean and R. Stoean, *Support Vector Machines and Evolutionary Algorithms for Classification*, Intelligent Systems Reference Library 69,
DOI: 10.1007/978-3-319-06941-8_8, © Springer International Publishing Switzerland 2014

As for comprehensibility, ESVMs are just more direct than SVMs in providing the weights. Runtime is increased by addressing the entire data set each time the evaluation of an individual takes place; chunking partly resolves it. Nevertheless, another advantage over SVMs is that ESVMs can be endowed with a GA feature selector, performed simultaneously with weight evolution. This extracts the more informative attributes, as a more understandable result of the classification label. The end-user can therefore view what are the indicators that influence the outcome.

- A sequential SVM-CC hybridization (Chap. 7) should be more comprising of the advantages of both. SVMs relabel the training data and the EA can this time address a noise-free collection to extract prototypes from. A shorter highly informative data set can be also obtained by referring only support vectors following the SVMs. Attribute thresholds are again generated by CC, as the better performing of the two presented EA approaches to classification. A HC is included again for synchronized feature selection. A pyramid of the importance of each problem variable for the triggered outcome can be additionally perceived. Finally, feature selection at the level of obtained prototypes presents a comprehensible format and image of those significant attributes for each class and the thresholds by which they differentiate between outcomes.

Having all these reviewed, can we now finally answer the question? Yes. When encountering a real-world problem, plain SVMs can give a first fast prediction. If this is not satisfactory, ESVMs can model the given data with non-traditional kernels. Once a good prediction is achieved, CC can extract the logic behind the black-box learning on the newly remodeled samples. Feature selection and hierarchy of attributes can be triggered along, as the EA can easily be endowed with many procedures for simultaneous information.

Or, in other words, we seem to have answered the question as Mark Twain would have put it:

> I was gratified to be able to answer promptly, and I did. I said I did not know.

References

Aguilar-Ruiz, J.S., Riquelme, J.C., Toro, M.: Evolutionary learning of hierarchical decision rules. IEEE Transactions on Systems, Man, and Cybernetics, Part B: Cybernetics 33(2), 324–331 (2003)

Akay, M.F.: Support vector machines combined with feature selection for breast cancer diagnosis. Expert Syst. Appl. 36(2), 3240–3247 (2009), http://dx.doi.org/10.1016/j.eswa.2008.01.009, doi:10.1016/j.eswa.2008.01.009

Bacardit, J., Butz, M.V.: Data mining in learning classifier systems: Comparing xcs with gassist. In: Kovacs, T., Llorà, X., Takadama, K., Lanzi, P.L., Stolzmann, W., Wilson, S.W. (eds.) Learning Classifier Systems. LNCS (LNAI), vol. 4399, pp. 282–290. Springer, Heidelberg (2007), http://dx.doi.org/10.1007/978-3-540-71231-2_19

Bache, K., Lichman, M.: UCI machine learning repository (2013), http://archive.ics.uci.edu/ml

Bäck, T.: Evolutionary Algorithms in Theory and Practice. Oxford University Press (1996)

Bäck, T., Fogel, D.B., Michalewicz, Z. (eds.): Handbook of Evolutionary Computation. Institute of Physics Publishing and Oxford University Press (1997)

Barakat, N., Diederich, J.: Eclectic rule-extraction from support vector machines. International Journal of Computational Intelligence 2(1), 59–62 (2005)

Barakat, N.H., Bradley, A.P.: Rule extraction from support vector machines: Measuring the explanation capability using the area under the roc curve. In: ICPR (2), pp. 812–815. IEEE Computer Society (2006)

Bartz-Beielstein, T.: Experimental Research in Evolutionary Computation – The New Experimentalism. Natural Computing Series. Springer, Berlin (2006)

Belciug, S., El-Darzi, E.: A partially connected neural network-based approach with application to breast cancer detection and recurrence. In: IEEE Conf. of Intelligent Systems, pp. 191–196. IEEE (2010)

Belciug, S., Gorunescu, F.: A hybrid neural network/genetic algorithm applied to breast cancer detection and recurrence. Expert Systems 30(3), 243–254 (2013)

Bessaou, M., Petrowski, A., Siarry, P.: Island model cooperating with speciation for multimodal optimization. In: Deb, K., Rudolph, G., Lutton, E., Merelo, J.J., Schoenauer, M., Schwefel, H.-P., Yao, X. (eds.) PPSN 2000. LNCS, vol. 1917, pp. 437–446. Springer, Heidelberg (2000), http://dblp.uni-trier.de/db/conf/ppsn/ppsn2000.html#BessaouPS00

Bojarczuk, C., Lopes, H., Freitas, A.: Genetic programming for knowledge discovery in chest-pain diagnosis. IEEE Engineering in Medicine and Biology Magazine 19(4), 38–44 (2000), doi:10.1109/51.853480

Bosch, R.A., Smith, J.A.: Separating hyperplanes and the authorship of the disputed federalist papers. Amer. Math. Month. 105(7), 601–608 (1998)

Boser, B.E., Guyon, I.M., Vapnik, V.: A training algorithm for optimal margin classifiers. In: Proceedings of the 5th Annual ACM Workshop on Computational Learning Theory, pp. 11–152 (1992)

Branke, J., Deb, K., Miettinen, K., Słowiński, R. (eds.): Multiobjective Optimization. LNCS, vol. 5252. Springer, Heidelberg (2008), http://books.google.ro/books?id=N-1hWMNUa2EC

Burges, C.J.C.: A tutorial on support vector machines for pattern recognition. Data Mining and Knowledge Discovery 2, 121–167 (1998)

Chang, C.C., Lin, C.J.: LIBSVM: a library for support vector machines (2001), http://www.csie.ntu.edu.tw/~cjlin/libsvm

Cherkassky, V., Mulier, F.M.: Learning from Data: Concepts, Theory, and Methods. Wiley (2007), http://books.google.ro/books?id=IMGzP-IIaKAC

Cioppa, A.D., De Stefano, C., Marcelli, A.: Where are the niches? dynamic fitness sharing. IEEE Transactions on Evolutionary Computation 11(4), 453–465 (2007)

Coello, C.C., Lamont, G., van Veldhuizen, D.: Evolutionary Algorithms for Solving Multi-Objective Problems. Genetic and Evolutionary Computation. Springer (2007), http://books.google.ro/books?id=rXIuAMw3lGAC

Cortes, C., Vapnik, V.: Support vector networks. J. Mach. Learn., 273–297 (1995)

Courant, R., Hilbert, D.: Methods of Mathematical Physics. Wiley Interscience (1970)

Cover, T.M.: Geometrical and statistical properties of systems of linear inequalities with applications in pattern recognition. IEEE Transactions on Electronic Computers EC-14, 326–334 (1965)

Cristianini, N., Shawe-Taylor, J.: An Introduction to Support Vector Machines. Cambridge University Press (2000)

de Jong, K.A.: An analysis of the behavior of a class of genetic adaptive systems. PhD thesis, University of Michigan, Ann Arbor (1975)

de Jong, K.A., Spears, W.M., Gordon, D.F.: Using genetic algorithms for concept learning. In: Grefenstette, J. (ed.) Genetic Algorithms for Machine Learning, pp. 5–32. Springer US (1994), http://dx.doi.org/10.1007/978-1-4615-2740-4_2, doi:10.1007/978-1-4615-2740-4_2

de Souza, B.F., de Leon, A.P., de Carvalho, F.: Gene selection based on multi-class support vector machines and genetic algorithms. Journal of Genetics and Molecular Research 4(3), 599–607 (2005)

Deb, K., Goldberg, D.E.: An investigation of niche and species formation in genetic function optimization. In: Proceedings of the Third International Conference on Genetic Algorithms, pp. 42–50. Morgan Kaufman, San Francisco (1989)

Diederich, J.: Rule extraction from support vector machines: An introduction. In: Diederich, J. (ed.) Rule Extraction from Support Vector Machines. SCI, vol. 80, pp. 3–31. Springer, Heidelberg (2008)

Dorigo, M., Stützle, T.: Ant colony optimization. MIT Press (2004)

Dumitrescu, D.: Genetic chromodynamics. Studia Universitatis Babes-Bolyai Cluj-Napoca, Ser Informatica 45(1), 39–50 (2000)

Dumitrescu, D., Lazzerini, B., Jain, L.C., Dumitrescu, A.: Evolutionary computation. CRC Press, Inc., Boca Raton (2000)

Eads, D., Hill, D., Davis, S., Perkins, S., Ma, J., Porter, R., Theiler, J.: Genetic algorithms and support vector machines for time series classification. In: Proc. Symposium on Optical Science and Technology, pp. 74–85 (2002)

Eiben, A.E., Smith, J.E.: Introduction to Evolutionary Computing. Springer, Berlin (2003)

Farquad, M.A.H., Ravi, V., Raju, S.B.: Support vector regression based hybrid rule extraction methods for forecasting. Expert Syst. Appl. 37(8), 5577–5589 (2010), http://dx.doi.org/10.1016/j.eswa.2010.02.055, doi:10.1016/j.eswa.2010.02.055

Fletcher, R.: Practical Methods of Optimization. Wiley (1987)

Fogel, D.B.: Evolutionary Computation. IEEE Press (1995)

Fogel, D.B.: Why Evolutionary Computation? In: Handbook of Evolutionary Computation. Oxford University Press (1997)

Freitas, A.A.: A genetic programming framework for two data mining tasks: Classification and generalized rule induction. In: Koza, J.R., Deb, K., Dorigo, M., Fogel, D.B., Garzon, M., Iba, H., Riolo, R.L. (eds.) Genetic Programming 1997: Proceedings of the Second Annual Conference, pp. 96–101. Morgan Kaufmann, Stanford University, CA, USA (1997), http://citeseer.nj.nec.com/43454.html

Friedrichs, F., Igel, C.: Evolutionary tuning of multiple svm parameters. In: Proc. 12th ESANN, pp. 519–524 (2004)

Fung, G., Sandilya, S., Bharat Rao, R.: Rule extraction from linear support vector machines. In: Proceedings of the Eleventh ACM SIGKDD International Conference on Knowledge Discovery in Data Mining, KDD 2005, pp. 32–40. ACM, New York (2005), http://doi.acm.org/10.1145/1081870.1081878, doi:10.1145/1081870.1081878

Giordana, A., Saitta, L., Zini, F.: Learning disjunctive concepts by means of genetic algorithms. In: Proceedings of the Eleventh International Conference on Machine Learning, pp. 96–104 (1994)

Goldberg, D.E.: Genetic Algorithms in Search, Optimization and Machine Learning. Addison-Wesley, Reading (1989)

Goldberg, D.E., Richardson, J.: Genetic algorithms with sharing for multimodal function optimization. In: Grefenstette, J.J. (ed.) Genetic Algorithms and Their Application, pp. 41–49. Lawrence Erlbaum, Hillsdale (1987)

Gorunescu, F.: Data Mining - Concepts, Models and Techniques. Intelligent Systems Reference Library, vol. 12. Springer (2011)

Gorunescu, F., Belciug, S.: Evolutionary strategy to develop learning-based decision systems. application to breast cancer and liver fibrosis stadialization. Journal of Biomedical Informatics (2014), http://dx.doi.org/10.1016/j.jbi.2014.02.001

Gorunescu, F., Gorunescu, M., Saftoiu, A., Vilmann, P., Belciug, S.: Competitive/collaborative neural computing system for medical diagnosis in pancreatic cancer detection. Expert Systems 28(1), 33–48 (2011)

Gorunescu, F., Belciug, S., Gorunescu, M., Badea, R.: Intelligent decision-making for liver fibrosis stadialization based on tandem feature selection and evolutionary-driven neural network. Expert Syst. Appl. 39(17), 12,824–12,832 (2012)

Hastie, T., Tibshirani, R., Friedman, J.: The Elements of Statistical Learning. Springer Series in Statistics. Springer New York Inc., New York (2001)

Haykin, S.: Neural Networks: A Comprehensive Foundation. Prentice Hall (1999)

Holland, J.H.: Adaptation in Natural and Artificial Systems. University of Michigan Press, Ann Arbor (1975)

Holland, J.H.: Escaping brittleness: The possibilities of general purpose learning algorithms applied to parallel rule-based systems. Machine Learning 2, 593–623 (1986)

Howley, T., Madden, M.G.: The genetic evolution of kernels for support vector machine classifiers. In: Proc. of 15th Irish Conference on Artificial Intelligence and Cognitive Science (2004), http://www.it.nuigalway.ie/m_madden/profile/pubs.html

Hsu, C.W., Lin, C.J.: A comparison of methods for multi-class support vector machines. IEEE Trans. NN 13(2), 415–425 (2004)

Huysmans, J., Baesens, B., Vanthienen, J.: Using rule extraction to improve the comprehensibility of predictive models. Tech. rep., KU Leuven (2006)

Joachims, T.: Text categorization with suport vector machines: Learning with many relevant features. In: Nédellec, C., Rouveirol, C. (eds.) ECML 1998. LNCS, vol. 1398, pp. 137–142. Springer, Heidelberg (1998)

Joachims, T.: SVM light (2002), http://svmlight.joachims.org

Johansson, U., Konig, R., Niklasson, L.: Genetic rule extraction optimizing brier score. In: Proceedings of the 12th Annual Conference on Genetic and Evolutionary Computation, GECCO 2010, pp. 1007–1014. ACM, New York (2010), http://doi.acm.org/10.1145/1830483.1830668, doi:10.1145/1830483.1830668

Jun, S.H., Oh, K.W.: An evolutionary statistical learning theory. Comput. Intell. 3(3), 249–256 (2006)

Kandaswamy, K.K., Pugalenthi, G., Moller, S., Hartmann, E., Kalies, K.U., Suganthan, P.N., Martinetz, T.: Prediction of apoptosis protein locations with genetic algorithms and support vector machines through a new mode of pseudo amino acid composition. Protein and Peptide Letters 17(12), 1473–1479 (2010), http://www.ingentaconnect.com/content/ben/ppl/2010/00000017/00000012/art00003

Kramer, O., Hein, T.: Stochastic feature selection in support vector machine based instrument recognition. In: Mertsching, B., Hund, M., Aziz, Z. (eds.) KI 2009. LNCS, vol. 5803, pp. 727–734. Springer, Heidelberg (2009), http://dx.doi.org/10.1007/978-3-642-04617-9_91

Langdon, W., Poli, R.: Foundations of Genetic Programming. Springer, Heidelberg (2001)

Li, J.P., Balazs, M.E., Parks, G.T., Clarkson, P.J.: A species conserving genetic algorithm for multimodal function optimization. Evolutionary Computation 10(3), 207–234 (2002)

Li, Q., Ong, A.Y., Suganthan, P.N., Thing, V.L.L.: A novel support vector machine approach to high entropy data fragment classification. In: Clarke, N.L., Furnell, S., von Solms, R. (eds.) SAISMC, pp. 236–247. University of Plymouth (2010)

Marinakis, Y., Dounias, G., Jantzen, J.: Pap smear diagnosis using a hybrid intelligent scheme focusing on genetic algorithm based feature selection and nearest neighbor classification. Comp. in Bio. and Med. 39(1), 69–78 (2009)

Markowska-Kaczmar, U., Chumieja, M.: Discovering the mysteries of neural networks. Int. J. Hybrid Intell. Syst. 1(3,4), 153–163 (2004), http://dl.acm.org/citation.cfm?id=1232820.1232824

Markowska-Kaczmar, U., Wnuk-Lipiński, P.: Rule extraction from neural network by genetic algorithm with pareto optimization. In: Rutkowski, L., Siekmann, J.H., Tadeusiewicz, R., Zadeh, L.A. (eds.) ICAISC 2004. LNCS (LNAI), vol. 3070, pp. 450–455. Springer, Heidelberg (2004)

Martens, D., Baesens, B., Gestel, T.V., Vanthienen, J.: Comprehensible credit scoring models using rule extraction from support vector machines. European Journal of Operational Research 183(3), 1466–1476 (2007)

Martens, D., Baesens, B., Gestel, T.V.: Decompositional rule extraction from support vector machines by active learning. IEEE Transactions on Knowledge and Data Engineering 21, 178–191 (2009), http://doi.ieeecomputersociety.org/10.1109/TKDE.2008.131

Mercer, J.: Functions of positive and negative type and their connection with the theory of integral equations. Transactions of the London Philosophical Society (A) 209, 415–446 (1908)

Michalewicz, Z.: Genetic Algorithms + Data Structures = Evolution Programs, 3rd edn. Springer, London (1996)

Mierswa, I.: Evolutionary learning with kernels: A generic solution for large margin problems. In: Proc. of the Genetic and Evolutionary Computation Conference, pp. 1553–1560 (2006a)

Mierswa, I.: Making indefinite kernel learning practical. Tech. rep., University of Dortmund (2006b)

Montgomery, D.C.: Design and Analysis of Experiments. John Wiley & Sons (2006)

Núñez, H., Angulo, C., Català, A.: Rule extraction from support vector machines. In: Verleysen, M. (ed.) ESANN, pp. 107–112 (2002)

Ozbakir, L., Baykasoglu, A., Kulluk, S., Yapici, H.: Taco-miner: An ant colony based algorithm for rule extraction from trained neural networks. Expert Syst. Appl. 36(10), 12,295–12,305 (2009), http://dx.doi.org/10.1016/j.eswa.2009.04.058, doi:10.1016/j.eswa.2009.04.058

Palmieri, F., Fiore, U., Castiglione, A., De Santis, A.: On the detection of card-sharing traffic through wavelet analysis and support vector machines. Appl. Soft Comput. 13(1), 615–627 (2013), http://dx.doi.org/10.1016/j.asoc.2012.08.045, doi:10.1016/j.asoc.2012.08.045

Perez-Cruz, F., Figueiras-Vidal, A.R., Artes-Rodriguez, A.: Double chunking for solving svms for very large datasets. In: Proceedings of Learning 2004. Elche, Spain Eprints (2004), pascal-network.org/archive/00001184/01/learn04.pdf

Pintea, C.M.: Advances in Bio-inspired Computing for Combinatorial Optimization Problems. Intelligent Systems Reference Library, vol. 57. Springer (2014)

Platt, J.C., Cristianini, N., Shawe-Taylor, J.: Large margin dags for multiclass classification. In: Proc. of Neural Information Processing Systems, pp. 547–553 (2000)

Potter, M.A., de Jong, K.A.: A cooperative coevolutionary approach to function optimization. In: Davidor, Y., Männer, R., Schwefel, H.-P. (eds.) PPSN 1994. LNCS, vol. 866, pp. 249–257. Springer, Heidelberg (1994)

Potter, M.A., Meeden, L.A., Schultz, A.C.: Heterogeneity in the coevolved behaviors of mobile robots: The emergence of specialists. In: Proceedings of the 17th International Conference on Artificial Intelligence, pp. 1337–1343 (2001)

Sarker, R., Mohammadian, M., Yao, X. (eds.): Evolutionary Optimization. Kluwer Academic Pusblishers (2002)

Schwefel, H.P.: Why Evolutionary Computation? In: Handbook of Evolutionary Computation. Oxford University Press (1997)

Schwefel, H.P., Wegener, I., Weinert, K. (eds.): Advances in Computational Intelligence. Springer (2003)

Shen, Q., Mei, Z., Ye, B.X.: Simultaneous genes and training samples selection by modified particle swarm optimization for gene expression data classification. Computers in Biology and Medicine 39(7), 646–649 (2009), http://www.sciencedirect.com/science/article/pii/S0010482509000833, doi:http://dx.doi.org/10.1016/j.compbiomed.2009.04.008

Shir, O., Bäck, T.: Niching in evolution strategies. In: Proceedings of Genetic and Evolutionary Computation Conference (GECCO 2005), pp. 915–916. ACM, New York (2005), doi:http://doi.acm.org/10.1145/1068009.1068162

Smith, S.F.: A learning system based on genetic adaptive algorithms. PhD thesis, University of Pittsburgh (Dissertation Abstracts International, 41, 4582B; University Microfilms No. 81-12638) (1980)

Stoean, C., Dumitrescu, D.: Elitist generational genetic chromodynamics as a learning classifier system. Annals of University of Craiova, Mathematics and Computer Science Series 33(1), 132–140 (2006)

Stoean, C., Stoean, R.: Evolution of cooperating classification rules with an archiving strategy to underpin collaboration. In: Teodorescu, H.-N., Watada, J., Jain, L.C. (eds.) Intelligent Systems and Technologies. SCI, vol. 217, pp. 47–65. Springer, Heidelberg (2009a)

Stoean, C., Stoean, R.: Evolution of cooperating classification rules with an archiving strategy to underpin collaboration. In: Teodorescu, H.-N., Watada, J., Jain, L.C. (eds.) Intelligent Systems and Technologies. SCI, vol. 217, pp. 47–65. Springer, Heidelberg (2009b)

Stoean, C., Stoean, R.: Post-evolution of class prototypes to unlock decision making within support vector machines. Applied Soft Computing (2013a) (submitted)

Stoean, C., Preuss, M., Gorunescu, R., Dumitrescu, D.: Elitist generational genetic chromodynamics - a new radii-based evolutionary algorithm for multimodal optimization. In: The 2005 IEEE Congress on Evolutionary Computation (CEC 2005), pp. 1839–1846 (2005)

Stoean, C., Preuss, M., Dumitrescu, D., Stoean, R.: Cooperative evolution of rules for classification. In: IEEE Post-Proceedings Symbolic and Numeric Algorithms for Scientific Computing 2006, pp. 317–322 (2006)

Stoean, C., Stoean, R., Lupsor, M., Stefanescu, H., Badea, R.: a) Feature selection for a cooperative coevolutionary classifier in liver fibrosis diagnosis. Comput. Biol. Med. 41(4), 238–246 (2011), http://dx.doi.org/10.1016/j.compbiomed.2011.02.006, doi:10.1016/j.compbiomed.2011.02.006

Stoean, R.: Evolutionary computation. application in data analysis and machine learning. PhD thesis, Babes-Bolyai University of Cluj-Napoca, Romania (2008)

Stoean, R., Stoean, C.: b) Modeling medical decision making by support vector machines, explaining by rules of evolutionary algorithms with feature selection. Expert Systems with Applications 40(7), 2677–2686 (2013), http://www.sciencedirect.com/science/article/pii/S0957417412012171, doi:http://dx.doi.org/10.1016/j.eswa.2012.11.007

Stoean, R., Preuss, M., Stoean, C., Dumitrescu, D.: Concerning the potential of evolutionary support vector machines. In: Proc. of the IEEE Congress on Evolutionary Computation, vol. 1, pp. 1436–1443 (2007)

Stoean, R., Preuss, M., Stoean, C., El-Darzi, E., Dumitrescu, D.: An evolutionary approximation for the coefficients of decision functions within a support vector machine learning strategy. In: Hassanien, A.E., Abraham, A., Vasilakos, A., Pedrycz, W. (eds.) Foundations of Computational, Intelligence. SCI, vol. 201, pp. 83–114. Springer, Heidelberg (2009), http://dx.doi.org/10.1007/978-3-642-01082-8_4

Stoean, R., Preuss, M., Stoean, C., El-Darzi, E., Dumitrescu, D.: Support vector machine learning with an evolutionary engine. Journal of the Operational Research Society 60(8), 1116–1122 (2009b)

Stoean, R., Stoean, C., Lupsor, M., Stefanescu, H., Badea, R.: Evolutionary-driven support vector machines for determining the degree of liver fibrosis in chronic hepatitis c. Artif. Intell. Med. 51, 53–65 (2011b), http://dx.doi.org/10.1016/j.artmed.2010.06.002, doi:http://dx.doi.org/10.1016/j.artmed.2010.06.002

Strumbelj, E., Bosnic, Z., Kononenko, I., Zakotnik, B., Kuhar, C.G.: Explanation and reliability of prediction models: the case of breast cancer recurrence. Knowl. Inf. Syst. 24(2), 305–324 (2010), http://dx.doi.org/10.1007/s10115-009-0244-9, doi:10.1007/s10115-009-0244-9

Vapnik, V.: Estimation of Dependences Based on Empirical Data. Springer (1982)

Vapnik, V.: Inductive principles of statistics and learning theory. Mathematical Perspectives on Neural Networks (1995a)

Vapnik, V.: The nature of statistical learning theory. Springer-Verlag New York, Inc., New York (1995b)

Vapnik, V.: Neural Networks for Intelligent Signal Processing. World Scientific Publishing (2003)

Vapnik, V., Chervonenkis, A.: Uniform convergence of frequencies of occurence of events to their probabilities. Dokl. Akad. Nauk SSSR 181, 915–918 (1968)

Vapnik, V., Chervonenkis, A.: Theorie der Zeichenerkennung. Akademie-Verlag (1974)

Venturini, G.: Sia: A supervised inductive algorithm with genetic search for learning attributes based concepts. In: Brazdil, P.B. (ed.) ECML 1993. LNCS, vol. 667, pp. 280–296. Springer, Heidelberg (1993), http://dx.doi.org/10.1007/3-540-56602-3_142

Wiegand, R.P.: Analysis of cooperative coevolutionary algorithms. PhD thesis, Department of Computer Science, George Mason University (2003)

Wiegand, R.P., Liles, W.C., de Jong, K.A.: An empirical analysis of collaboration methods in cooperative coevolutionary algorithms. In: Proceedings of GECCO 2001, pp. 1235–1245 (2001)

Wilson, S.W.: Classifier fitness based on accuracy. Evol. Comput. 3(2), 149–175 (1995), http://dx.doi.org/10.1162/evco.1995.3.2.149, doi:10.1162/evco.1995.3.2.149

Index

Printed in the United States
By Bookmasters